EQUIVALENT CIRCUIT MODEL OF

QUANTUM MECHANICS

EQUIVALENT CIRCUIT MODEL OF

QUANTUM MECHANICS

MASAKAZU SHOJI

EQUIVALENT CIRCUIT MODEL OF QUANTUM MECHANICS

iUniverse books may be ordered through booksellers or by contacting:

iUniverse
1663 Liberty Drive
Bloomington, IN 47403
www.iuniverse.com
844-349-9409

ISBN: 978-1-6632-4894-7 (sc)
ISBN: 978-1-6632-4896-1 (hc)
ISBN: 978-1-6632-4895-4 (e)

Library of Congress Control Number: 2022922975

Print information available on the last page.

iUniverse rev. date: 02/02/2023

CONTENTS

PREFACE

Physics students have many basic questions in their quantum mechanics class. From my remote physics student days in the 1950s, I had many difficulties in understanding the basic concepts of quantum mechanics. They appeared almost mysteries to me, and many such mysteries have still not been plainly explained. The theory works as if by magic for trained physicists, but for me, working in the physics' peripheral area, I feel as if the theory seems to have no solid foundation for understanding.

This work summarizes my effort at trying to understand some of such *quantum mysteries.* I show what I believe is the possible explanation of some of the mysteries, at least, to myself. The interpretation that was convincing to me may not appear acceptable for everyone. Yet I believe that the explanation, of which I struggled to convince myself, has some seed of truth in it. This work is a follow-up of what I published in the two previous books, *Dynamics of Digital Excitation*, published by Kluwer Academic in 1997, and in the last chapter, "Sense of mystery," in *Self-Consciousness, Human Brain as Data Processor*, published by iUniverse in 2020. The first book was purely on electronics and physics, but the second book dealt with psychology, in which I used quantum phenomena as an example of mysteries of the human mind, and I tried to combat the now popular belief that human self-consciousness is a quantum phenomenon, which I believe is entirely wrong.

I approach the quantum mysteries from viewpoints that differ from those found in the authoritative textbooks. In my work, I am well aware that the boundary between rigorous logical theory and rational analogy is not clear. In spite of that, I try to present as many comprehensible explanations as possible by creating models of quantum phenomena, so that the basic laws of nature do not remain as mysterious as they are now. To do so, I adopt different viewpoints about the origin of quantum mechanical phenomena.

First, I do not repeat the origin of quantum particles and forces to the phase transition of the early universe. Instead, I consider that the present quantum laws retain features that emerged at various developmental phases of the early universe, starting from the point singularity. As the universe's size increased, some basic features remained, but various new features emerged, added to the older features, and remained to the present. In effect, I introduce a historical viewpoint in the quantum physics laws. As such, what appear to be the simplest and the most mysterious features of quantum mechanics are the most archaic.

The second viewpoint is a search for the models of quantum phenomena that emerged during the various phases of the universe's early development within the classical and the elementary quantum physics domain. Wave mechanical representation of a particle's probability and quantization of parameters are the two basic features of quantum mechanics. In the classical domain, there are models that accommodate these two features. They are the equivalent circuit models in modern electronics. The analog and digital equivalent circuit models reflect the various developmental stages of the universe.

So I sought for various equivalent circuit models suitable to explain mysterious quantum phenomena. Following the historical development of the quantum laws, I developed the proper equivalent circuit models representing the development of the universe. Yet I needed to expand the concept of the basic equivalent circuit models including various exotic modes of operations such as self-altering circuits. This is perhaps the most unique feature of my work; this idea emerged from my background as electronic circuit theorist and systems engineer in my active profession.

There is another interesting feature of my model-based approach, which is one basic reason of my present equivalent circuit–based work. Some strange *qualitative* features of quantum world emerged from the model I introduced in this book, based on the simple assumptions I made. To reach the same conclusion, the standard model-less quantum theory must go through complex or unusual mathematics. This is the reason why I have been fascinated by this work. The first concerns the two kinds of elementary particles, fermion and boson, that are derived in the standard quantum mechanics textbooks from a curious algebra of symmetrized and anti-symmetrized wave functions. The second is that particles' internal symmetry conversion (fermion versus boson) makes fermions move in space. The third is the explanation of the mechanism of an elementary particle's spin, that also emerges from the model, without going through relativistic mathematics. The fourth is a simple explanation of the mysterious phenomenon of quantum entanglement. These are the reasons why I have labored to refine the crude original idea of quantum mechanics I published before. I believe that quantum physics can be reformulated on some kind of models, as classical physics has been. Physics must be a model-based science.

The third viewpoint is that I reexamined the original concept of atoms from the Greek philosophers, that elementary particles are the final products of splitting a macroscopic object to the limit. The splitting is not limited to materials but includes information. By splitting the minimum information, one bit, I get probability, whose *segments* distributed over space and time develop quantum phenomena. The probability segments behave like diffusing particles, and they follow the semiclassical laws of diffusion, subject to proper generalization and reinterpretation.

The effort of trying to make models of quantum effects is not the mainstream activity of quantum research at present. Indeed, I admit that such a work is considered as a heresy in the mainstream of modern physics. I dare to step into this unpopular territory by my own *idle yet genuine curiosity.* The quantum world is filled with mysteries. Any mystery is a burden, and also is a challenge to the human mind. Explanation by a model of any mysterious phenomenon removes the mind's stresses. The human mind deals with the world that works basically according to classical physics. Any feature of explanation that appeals to the human mind must satisfy this basic character of the human mind. Then, the model must be understood by human mind in terms of its classical physics features. What are the differences between quantum and classical physics features? Here I need to think over the quantum mysteries from psychological viewpoints. Psychologically, the boundary between the classical and the quantum features is not so clear as has been believed, especially if the difference is observed from the science of human mind, psychology. There are many *quantum phenomena* in the classical physics domain.

This book primarily deals with the subjects in the boundary region between the quantum physics and the emerging new area of digital electronic circuit theory. I expect that not many readers are familiar to the both areas. I did not have an easy choice in the writing style. I had a choice to explain every basic concept in elementary details, but after excruciating struggle in my mind, I decided to make the text as compact as possible, so that readers can skim the basic idea without going too much into technical details. I must expect that readers have some background in the both areas, especially in electronic circuit theory and in elementary quantum mechanics—for instance, college students majoring in electronics engineering whose minor is physics. I expect that there are many students of this kind. My hope is that readers understand my intention. I hope that they get the basic message, that models of quantum mechanics can actually be built by using the electronic equivalent circuit model. Relying on the model, there are really not so many mysteries in the elementary quantum world. This is the basic message of this book.

This work is a continuation and summary of my two previous works referenced above. I owe a lot to my wife Marika, a psychologist, who helped during the time of this research through encouragement and support to dig into this unpopular yet unexplored boundary region between physics, electronics, and psychology.

1

PROBABILITY IN THE QUANTUM WORLD

1.01 Introduction

In quantum physics, definite physical parameters are displaced by probabilities. Why does this feature emerge? The basic approach of any natural science is reductionism; that is, splitting any complex object into its components, working with the simplified components, and then assembling the gained information to the comprehensible original object's attributes.

This approach reaches the limit when the object becomes the elementary particle that cannot be physically split anymore. Such an object is characterized by its unique parameters of mass, energy, momentum, and several universal *characters*, such as charge, spin, color, etc. To study the nature of elementary particles, some of the parameters must be further split, not physically but conceptually, into the components making them up. Here the concept of probability emerges. The physically undividable particle's parameters are conceptually divided into probabilities distributed over space and time; their relations and development are mathematically analyzed, and then assembled to understand the nature of the object. Here, division of the parameters and their assembly of the physically meaningful form are the two basic processes.

When any physical object is split into the level of elementary particles, the parameters are specified by several binary numbers, by using the actually measured parameter values as the unit. Electron mass is 1 bit, that is physically $9.11 x 10^{-31}$ kg in the conventional unit. Similarly the electronic charge is $1.60 x 10^{-19}$ coulomb, and the spin is $0.53 x 10^{-34}$ joule-sec. When these quantities are actually measured, the magnitude is represented in the theory by the basic unit's integral multiples. Then, no real fractional mass, spin, or charge is considered. Some such binary parameters are instead distributed probabilistically over space and time. This is what I call *splitting* of the basic information. The probability is not measurable by a single observation. The theory and experiment become only statistically comparable. What we measure are the statistical expectation values of the parameters.

This feature suggests that the transition from definite classical parameters to probabilities is the ultimate step of going down into the depths of reality by the reductionist approach; probabilistic description of phenomena cannot reach into any lower level. There is no probability of probability. Quantum physics has reached the ultimate depth of reality.

The probability of the elementary particle's parameters reveals the feature of randomness, which is reconstructed into the reality by the generalized diffusion process. This is the second step of assembling to understand the reality. In nature, mixing and randomization followed by the orderly state is executed always by some form of a diffusion process. The diffusion mechanism extracts integrated order from random component events. This feature manifests itself in the basic equation of quantum mechanics, the Schrödinger equation. This equation has a mathematical structure of the diffusion equation, and it displays various features of the common diffusion process, starting from the probabilistic basic data, and the emerging macroscopic order. Yet the diffusion process described by the classical diffusion equation is not general enough to cover every feature of the probabilistic phenomena of the quantum world. The concept of diffusion must be generalized to cover the entire field of quantum probability, including the awkward process of *probability field collapse*, to extract the reality. My discussion of quantum mechanics model begins with observation of the generalized diffusion process in quantum probability fields.

1.02 Probability

The states of most quantum systems are specified by probability. Probability is a concept not limited to quantum mechanics. There are many familiar cases of probabilistic experience in our secular life. I believe that probability in quantum mechanics is essentially the same as that experienced in everyday life, such as the lottery. The only difference is that purposeful human intervention, such as setting the number of participants or the number of winners, is involved in setting up and executing secular probabilities. Then, by considering the non-humanized, ideal cases of secular probability, I can get an insight into the probability of quantum mechanics.

In order to see the similarity between the lottery probability and the quantum probability, I consider the most simplified case of a lottery promoted in an extended area such as a country. There are N participants, where N is idealized to be a very large number, and there is only one winner. The participants are identified by number n, where n is between 1 and N. In this model, what is meant by the probability? Unit information, 1 bit, assigned to the winner, is split into N fragments of $1/N$ bit each. Winning is defined by assembling the N fragments of $1/N$ bit each to the winner's 1 bit somewhere in the N locations of the country. Each of the N participants holds a fraction of the 1 bit, and the sum total of them is unity before decision. This is the *expectation value* of the participants. At the time of decision making, N fragments of $1/N$ bit each accumulate to a single location to reconstruct the winner's 1 bit.

I compare this lottery probability with the quantum probability of a single particle in an extended spatial area. Each location carries a probability fragment of finding the particle. At the time of observation, the probability fragments accumulate *instantly* at a single location, where the particle is detected. In the later sections, I call the probability fragment a probability *segment*, and I study the probability segment's dynamics.

The lottery machine has balls, each of which carries the identification number of a participant. The lottery ticket is sold in the country, not randomly, but in some regularity; for instance, the participant's identification number increases from the east to the west of the country. The lottery machine is a jar, in which the balls carrying the participant's numbers are put in from 1 to N in that order. At the beginning, the locations of the participants in the country and the locations of their balls in the jar have a well-defined mapping relation. That is, a ball's location in the lottery machine is continuously *mapped* to the participant's location in the country. This is the initial condition, common to both the lottery operation and the quantum particle in the probability field.

In the lottery machine, after it is filled with the balls, air is blown in from the bottom. The balls execute convective mixing in the jar, which is a diffusion process, starting from the known initial condition of the packed balls. At the top of the jar, there is a pipe and pump arrangement that sucks up a single ball at the time of decision making. The ball's number determines the winner. The chance of an individual participating in the lottery winning is not exactly the same for all. The nonuniformity is created by the initial conditions and the nonideal aspects of the lottery machine; for instance, each ball may have slightly different weight that the participant cannot control, and the wall of the mixing jar is not perfectly smooth. Such features create nonuniformity of the participant's chance of winning. Similarly, the probability of the quantum particle's location is subjected to the potential energy in the quantum probability field. The particle is likely to stay at the low potential energy area more frequently than the high potential area. The particle's location $\mathbf{r}$ (vector, boldface character) is equivalent to the location of the ball carrying the identification number n in the lottery machine jar. The probability of participant n at location $\mathbf{r}$ in the country at time t, $P(\mathbf{r},t)$ is determined by how close his ball is to the top pipe of the lottery machine at the time t. The processes of a lottery and of quantum particle location have exactly the same interpretation, and the development, if $P(\mathbf{r},t)$, is interpreted as the probability of finding the particle at time t at location $\mathbf{r}$.

In both cases, if the probabilistic processes are compared, the probability distribution in the field changes continuously with time t and location (n or $\mathbf{r}$). The lottery and quantum probabilities' similarity improves if the lottery participants number N is so large that it can be regarded practically as infinity, and they are distributed continuously over the country. The balls in the lottery machine simulate the diffusion process of the quantum probability segments subjected to the potential energy and the initial condition. The lottery's chance of winning referred to location $\boldsymbol{r}$ in the country at time t is a continuous function $P(\boldsymbol{r},t)$. As long as the probability setting is not disturbed, $P(\boldsymbol{r},t)$ is a continuous function of the spatial coordinate $\boldsymbol{r}$ and time t. The quantum probability has a similar function $P(\boldsymbol{r},t)$, and that is determined by the Schrödinger equation, which provides the particle's continuous probability profile development. The similar functional relation exists between the probability of the lottery participants and the quantum particle; that is, a continuous mixing process by diffusion by *convection* changes the participant's probability. This similarity is valid, unless the number of participants of the lottery is small. Then, the continuous probability function $P(\boldsymbol{r},t)$ cannot be defined.

1.03 Diffusion of Probability Segments

When a pair of water colors is poured into the same container, they mix and create an intermediate color. This is by a process of mutual diffusion of the dye's particles interpenetrating into each other's space, thereby creating a uniform intermediate color. If this process is mathematically formulated, the microscopic flow of the dye particles is described by the diffusion equation. The flow rate of dye particle A from location 1 to 2 is proportional to the density difference of the dye particles A at location 1 and 2. The flow rate is given by $D_A(\psi_1 - \psi_2)/\Delta$, where ψ_1 and ψ_2 are the densities of dye A particles at location 1 and 2, D_A is the diffusion coefficient of dye A particle, and Δ is the distance between location 1 and 2. A similar relation holds for the flow of dye B particle.

This apparently simple mathematical description contains some special features of the diffusion mechanism. One such feature is the definition of the densities of dyes A and B. To derive the conventional diffusion equation, the volume elements in which dye A and B particles' densities are defined are infinitesimally small, but still contain a large number of dye particles. Here emerges a special case. If the number of dye particle decreases, ultimately to a few to only one or zero in the volume elements, the diffusion process creates microscopic randomness, along with overall macroscopic equalization of densities of the particles. The diffusion process bridges orderly and disorderly states. Macroscopic order emerges from microscopic disorder. This feature is relevant to the probabilistic interpretation of quantum phenomena.

Along with the size of the volume to define ψ_1 and ψ_2, the distance Δ between the volume elements matters. In the continuous mixing of water color, the distance can be set at infinitesimally small also, so that the diffusion process can be described by the partial differential equation. Yet if we consider the diffusion process in general, including the diffusion process of lottery probability segments, there is no reason why the distance needs to be infinitesimally short. Some significantly distant locations 1 and 2 may be involved in the diffusion mixing process. This point is clear if we consider the mixing process of balls in the lottery machine of the last section, when the number of participants is scattered over the country. In quantum mechanics, this feature is significant to consider the collapse of the probability field of the wave function by the same diffusion process, when realistic information is extracted by observation of the probability field. I believe that if the quantum probability field develops by a diffusion process, its collapse should also be described by the diffusion process, defined in more general terms.

In a typical situation in which the diffusion process determines the system's development, the diffusion coefficient D_A is positive and finite. Then the particles flow from the high-density location to the low-density location, thereby equalizing the density everywhere. When the probability profile is developing continuously without any external intervention, the diffusion coefficient may take a positive value or may even take a moderately negative value for a limited period of time. The effect establishes equilibrium, except when there is some other physical mechanism such as the system's phase transition. Mathematically, however, there is no reason why D_A cannot take a negative value or an extreme value, such as negative or positive infinity. If D_A is positive and infinitely large, the density equilibrium is established instantly. This effect violates the relativistic locality requirement (i.e., if something happens at location A, location B is affected only after the time defined by distance AB divided by the light velocity). The diffusion process is not compatible with relativity as it is simply defined.

If D_A is negative, the uneven feature of the density (or probability) profile increases; the high-density area increases, and the low-density area decreases. In the diffusion process of probability segments, if D_A is negative, the probability profile changes to reveal peak and valley features of the field. Since probability is limited to the range between 0 and 1, the mathematical procedure must take this into account. That is, the mathematics is not *analytic* all the way in time in order to suppress negative probability.

In generalizing the diffusion mechanism, the two effects I discussed must be included. That is, the diffusion current is able to flow between two points some distance apart, and the diffusion coefficient may take negative or infinite value. This happens at the time of reality extraction from the probability profile. D_A tends to negative infinity, and the probability segments flow some long distance apart. This is also the case in the change of the rule of probabilistic operation, such as that of a lottery. Given any effect, such as a lottery participant giving up the winning chance, the entire probability field changes instantly. Generally, any effect that externally affects the probability field's natural development is the cause of the long-distance interaction and a negative or infinite diffusion coefficient. Infinite speed probability fragment movement emerges in such cases. A long-distance interaction and an infinite diffusion coefficient both cause violation of the relativistic locality requirement.

In the case of a single winner of a lottery, or a single particle in a quantum field, the probability profile changes to a single peak of unity height at the time of observation. The distributed probability segments accumulate to the single winner's or single particle's location instantly. This process occurs by infinite velocity movement of the probability segments in the probability field. The infinite velocity probability setting signal emerges when the field is interrupted or observed; then the probability converts to reality. This is the reason why nonlocal features of the probability field emerge at the time of observation. In quantum mechanics, probability field observation is called the *collapse* of the quantum probability field set up by the wave function. Thus, the entire process of probabilistic development of quantum mechanics should be covered by the generalized diffusion process. This feature is elaborated in the next section.

1.04 Probability Field Collapse

The probability field that describes a quantum state is a generalized diffusion field. The diffusion field develops continuously, if there is no external intervention. If the field is subjected to intervention such as observation, the field changes instantly to reality. I believe that the continuous development and the abrupt change or decay of the probability field are both covered by the *generalized* diffusion process. The generalized diffusion process is a natural and mathematical extension of the continuous diffusion process and covers both quantum and everyday probability processes.

The parameters that characterize the generalized probability segment's diffusion process are the distance of movement of the probability segments and the sign and magnitude of the diffusion coefficient. In the conventional continuous development of diffusion, the probability segment's flow **J** (vector) from location 1 to 2 is given by the equation

$$\mathbf{J} = D_A[P(1) - P(2)]\mathbf{n}/\Delta$$

where **n** is the unit vector directed from location 1 to 2, and Δ is the distance between locations 1 and 2. The small volume surrounding location ***r*** defines the probability segment. The rate of change of the probability field at time t and location **r**, $P(\mathbf{r},t)$, satisfies the conservation equation

$$\frac{\partial P(\mathbf{r},t)}{\partial t} + \operatorname{div} \mathbf{J} = 0$$

The normalization of the probability is assured; that is, the sum total of the probability must be maintained at unity. In the continuous diffusion process, locations 1 and 2 are infinitesimally close to each other, and the diffusion coefficient D_A is either positive, zero, or only temporarily negative.

To define the generalized diffusion process, the two restrictions must be removed. Let us consider the case of only one particle in the quantum probability field. When the probability field is affected by the external intervention of observation, the overall feature of the entire probability field is revealed as reality. For that purpose, all the scattered probability segments at different locations must gather at a single location. For the process to take place, the partial differential equation describing the conventional diffusion process is not adequate. The algorithm cannot be described *analytically* by a differential equation, since interaction at not infinitesimally small distance is involved. The probability field must be converted to discrete cells, each of which contains probability segments $P(*)$. The operation exercised in the generalized diffusion process to these cells is as follows. The following numerical process is repeated to every pair of cells such as locations 1 and 2 in the probability field, in sequence. For locations 1 and 2, where Δ is the distance between the two locations, if $P(1) > P(2)$,

$P(1) \rightarrow p(1) + \delta P(1)$ and $P(2) \rightarrow P(2)\text{-}\delta P(1)$,

where $\delta P(1) = D[(P1) - P(2)]/\,\Delta$

and if $P(1)<P(2)$,

$(2) \rightarrow P(2) + \delta P(2)$ and $P(1) \rightarrow P(1) - \delta P(2)$,

where $\delta P(2) = D[P(2) - P(1)]/\,\Delta$

As for the *diffusion coefficient* D, it takes the large *positive* value in this numerical process, approximating $D \rightarrow \infty$ This process is executed subject to the condition that any minimum $P(*)$ is not negative, until $P(*) \rightarrow 1$ is reached at some location, where the particle is found and the operation is terminated. This means that the location of high probability gains probability segments and the location of low probability loses probability segments. Yet all the probability segments $P(*)$ are limited within the range of 0 and 1 during the operation.

In this generalized diffusion process, the distance between locations 1 and 2 is not infinitesimally small. This means that at the time of observation, the probability segments move the distance at infinite velocity, because Δ is not infinitesimally small and D is large enough to approximate infinity. This is possible, since probability segments carry no information. The probability is never carried by mass or energy, and therefore the relativity limit does not apply. Any information is carried by mass or energy, to tick the receiver, which requires energy. Probability signals need not do that.

Since the probability field contains only one particle, the particle emerges, most likely, at the probability field's location where the probability was originally the highest, but that is never guaranteed. A peculiar feature of this process is that the order of execution of the numerical operation matters, but the order is never specified by any means. Nature takes whatever order it actually chooses. Depending on the order of execution, the particle may emerge at an unpredictable location. For instance, if the probability field right before the observation has two equally high probability peaks at locations X and Y, which of X or Y ultimately displays the particle depends on the order of execution of the probability segment's movement. The order is not specified by any externally controlled mechanism; the order is probabilistically set by nature. This is one of the causes of the uncertainty emerging in quantum measurement. I suspect that the cause of this uncertainty is the fluctuation of the quantum vacuum, which generates and annihilates the virtual particle spontaneously and randomly at any given location. Yet I cannot specify the exact mechanism at this time.

Another peculiar feature is that the dynamics of the model of the probability field must have two different signal propagation mechanisms; one satisfies the locality condition. If the system is not observed or not disturbed, the system develops continuously, and the probability segment's movement is restricted by finite velocity. The other mechanism must allow infinite velocity probability setting signal propagation, which emerges when the probability field is externally interfered with by observation, or when the field's structure is changed by external intervention. I stress that this is a general feature of the probability field, including even everyday probabilistic operations such as a lottery. In my work, I use a single term *signal* to indicate both information or probability flow mechanisms.

1.05 Probability Signal Transmission

To develop the phenomena in a probability field, I need two types of signal transmission mechanisms from the discussions of the last section. The first mechanism transmits the information-carrying signal at finite, less than light velocity, and the second mechanism transmits the probability-setting signal instantly. The first mechanism is operative to develop the state of the system, and the second mechanism operates if the structure of the probability field is interrupted from the outside, by observation or by any other system's structural change. The two transmission mechanisms must have similar physical mechanisms, so that one can be related to the other naturally (see section 3.05).

To make a model of the signal transmission, no classical mechanical model is adequate. Any classical mechanical structure consists of the parts carrying mass. The structure cannot transmit a signal instantly. The model that provides both signal transmission mechanisms is the generalized and idealized linear and digital electronic equivalent circuit model. Such a linear circuit model transmits a quantum probability wave along a transmission line. This circuit must be linear, in order to satisfy the superposition principle of the probabilistic quantum states. The superposed quantum state carries the probability of existence of the particle propagating along the path. In this circuit model, the probability must be maintained within the range of 0 and 1. This means that each wave amplitude must neither diverge to infinity nor converge to zero. For that purpose, a lossless LC transmission line is the intuitive candidate, but that is not the proper model, because it does not represent the physics of quantum mechanics. The equation describing an LC transmission line does not match the basic quantum equation.

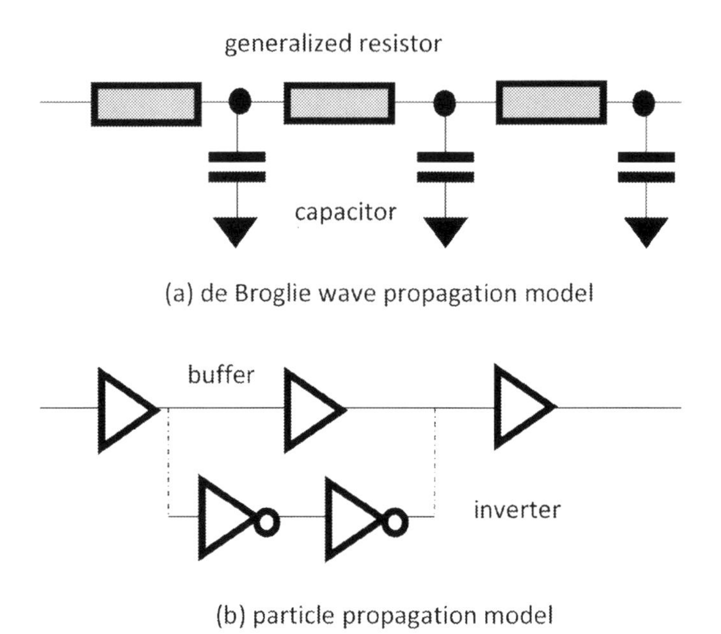

Figure 1.05.1 Two types of signal propagation

There is an interesting *historical reason*. Each physical phenomenon has its origin in the universe's history, and the history shows up if we consider the order of emergence of the circuit components making up the model. An LC circuit model requires inductance, and inductance's historical origin is not so old as the mechanism of quantum phenomena, as I discuss in sections 1.06 and 1.07. I believe that quantum laws retain the features of the state of the universe prior to the emergence of inductance, in which instant action at a distance dominated. The emergence of inductance introduced relativistic effects in the physical system. The instant action has been taken over by the relativistic effect of locality action; that is, the cause at location A creates the effect at location B at a later time, given by distance A–B divided by the light velocity. By the emergence of inductance in the equivalent circuit model in the expanding universe, some probabilistic parameters were replaced by definite parameters, i.e., information. Information transmission must satisfy locality requirements. Light velocity never set the limit of any phenomena during the earlier phases of the universe, when the parameters were all probabilistic.

Then, what are the models of the particle propagation? Figure 1.05.1(a) shows the RC ladder circuit of quantum (de Broglie) probability wave propagation. The resistor in the ladder circuit is purely imaginary to maintain the wave amplitude. If the resistor had a real value, the wave amplitude would diverge or converge to zero, depending on negative or positive resistor values, respectively. An imaginary resistor cannot actually be made, but the equivalent circuit model accommodates such components.

The node voltage of this ladder circuit is not externally observable, because the wave that propagates on the imaginary-RC chain circuit has complex number amplitude. The probability is given by the wave amplitude squared, which becomes an observable real number. Observation of the wave amplitude affects the model's operation. This is the process of extraction of reality from the probabilistic state of the circuit model. This process requires an infinite velocity signal propagation mechanism to set up a consistent model.

The digital circuit that propagates the particle's signal is a cascaded chain of digital buffers, each of which consists of two-stage cascaded inverters as shown in Figure 1.05.1(b). The first-stage inverter of the buffer is not capacitively loaded to maintain the transmitted waveforms intact (this stage works as the wave shape restorer). The second-stage inverter is the output stage of the buffer, which *can* be capacitively loaded to model real particle propagation. Since particle propagation must be observable, all the circuit parameters must have real values, and the circuit's observability depends on the node's capacitive loading (see section 2.03). This is because the measurement device requires extracting some energy from the observed system: in this case, from the capacitor of the circuit node. If there is no capacitor, there is no energy. Such a node is not observable. The energy stored in the capacitor models the substance of the particle (the mass or energy). A fully capacitively loaded node is observable, and I find a particle there. Whether the node is capacitively loaded or not determines whether or not the state of the particle can be observable. If it is observable, the full particle is observed. The observer gets the full substance of the particle. If the node is only partially loaded by the capacitor, the particles are observed only probabilistically. The particle propagates through the buffer chain at a velocity less than light velocity to satisfy the locality requirement of relativity.

The buffer chain models both particle and probability signal propagation, depending on its capacitive loading. If the buffer output node is not capacitively loaded, there is no substance of the particle, and the node is not observable. A signal propagates along the capacitively unloaded buffer section at infinite velocity. This is the infinitely fast probability signal propagation path. Such a transmission structure emerges and operates only when the probability field alters by external access, either by observation to extract reality, or by external intervention, affecting the structure of the probability field . As I show later, a digital buffer chain model is converted to such a structure from the source to the moving particle, and the only role of the chain is to transmit a probability signal to erase the particle's track. The buffer chain node in such a case is never externally observable, because that interferes with the model's operation (section 2.03). Using the pair of the linear and the digital equivalent circuit models, quantum phenomena can be properly described. Since they are both electronic circuit models, they can be naturally correlated each other in the physical mechanism (section 3.05).

It might appear curious that the linear ladder circuit model contains physically not realizable imaginary value resistors, and the digital chain model consists entirely of real circuit components. A two-stage cascaded buffer loaded with a capacitor reproduces the same real signal at the output, so it models real particle propagation. If it is not capacitively loaded, it is not observable, but it becomes the chain of infinitely fast probability signal propagation. This operation is to affect the particle state externally, such as by observation. The linear ladder circuit reproduces the same amplitude wave from one node to the next node. Many waves are superposed to make a wave packet that represents the particle. Then the linear and digital circuit models display exactly the same quantum phenomena of linear propagation of the particle, if the linear circuit's voltage amplitude squared is interpreted as the particle's existence probability, and the digital buffer chain node voltage is interpreted as the probability of finding the particle at the nodes. If I consider this feature in detail, a particle state consists of multiple linear ladder models. Details of this feature are discussed later in sections 2.09, 2.10, and 3.05, where I define the effective node voltage of the digital buffer chain model as the particle's existence probability. Since the two are both electronic equivalent circuit models, their relation can be naturally correlated in modeling the quantum phenomena.

As for the model of digital circuit technology, I use the presently most widely used CMOS technology, and I idealize and modify some of its features. I have worked out the theory of CMOS digital circuit technology (*Theory of CMOS Digital Circuits and Circuit Failures*, Princeton University Press, 1992). This technology has the advantage of node pull--up and pull-down symmetry (the circuit is symmetrical with respect to the pair of power supply buses), and that is proper to display particle and antiparticle symmetry in the quantum world. The other advantage is the capability to float the output node of the buffer to represent an indeterministic quantum state (section 3.11). As for the basics of the CMOS technology, there are many textbooks available. One of them is my own (*CMOS Digital Circuit Technology*, Prentice Hall, 1987).

1.06 Probabilistic Quantum System

The origin of quantum phenomena goes back to the earliest age of the universe. I see this *antique feature* in the quantum mechanics laws and in the model of the theory to describe the *present* quantum state. In order to understand the physical meaning of elementary quantum phenomena, I need to reflect the state of the universe's early history in the quantum effect's model. At the beginning, the size of the universe was so small and the materials were so densely packed that nothing had definite structure and mutual relation, such as the *quantum foam* (K. W. Ford, *The Quantum World*, Harvard University Press, 2004) in its archaic state. Because of the lack of structure and order, the object's properties could not be specified by any definite parameters, but only by probabilities. This feature is inherited to the present quantum phenomena displayed by elementary particles.

To describe the particles of the present quantum world, the models specifying their active (dynamic) and passive (static) states are required. What would be the proper model? Classical mechanical models are inflexible to deal with the probabilistic description. I suspect that there was a historical reason why quantum mechanics was developed without relying on models, but only on mathematics. In the early days of quantum mechanics development in the 1920s, electronic circuit theory was in its infancy, and the equivalent circuit models were never considered by physicists. Now that electronic circuit theory has been well developed, I can show that the proper model of the quantum world is the equivalent circuit model, which can be modified and adapted to describe quantum phenomena. Unfortunately, at present, not many electronics experts try to stretch their imagination into quantum physics. This is because of the strict institutionalization and the tight academic discipline in the engineering community, which unfortunately discourage free imagination in the unexplored boundary region beyond the established technical framework.

The equivalent circuit model deals with any quantitative (linear) and qualitative (digital) physical phenomena and can represent the modeled object without going too much into its structural details such as the object's shape, size, etc. By considering the ultimate limit of the equivalent circuit model in the exotic quantum physical limit, I recognized an insight into the state of the early universe, and its descendent, the present quantum phenomena, can be consistently understood.

I state my basic beliefs in the present work here. They are, *first*, any physical phenomena, quantum or classical, can be modeled and understood by setting up and analyzing the proper equivalent circuit model; *second*, the present quantum physics laws descended from the state of the ancient universe as they were then, to the present quantum laws; and *third*, the equivalent circuit model represents the functions of all the connected structures of the elements at any time of the universe's history. Here, the circuit model as it is conventional at present in the engineering community is not adequate, but it can be modified, expanded, and adapted to represent the quantum phenomena. I work out the model's details based on the three beliefs.

In order to model quantum or classical physics phenomena, the equivalent circuit model structures and parameters must be specified freely, by ignoring practical realizability. The model circuit must be small in its size, reflecting the quantum objects realistically, since the size of the equivalent circuit brings essentially new insights into the quantum physics. The self-altering circuit (which has not yet been studied in detail) must be included as its crucial element. The environment of the equivalent circuit model must be properly specified. Otherwise the circuit model should work in the conventional way, following Kirchhoff's laws.

The sizes of the elementary particles as presently considered place the requisite size of equivalent circuit model structures close to the limit of the spatial resolution set by general relativity. Their action times are also near the limit of the time resolution set up by the circuit's operating conditions. I leave open, at this moment, the question whether general relativity is relevant to describe the beginning of the universe, but the theory provides definite limit of the size of the circuit model that I can use. The limit is reached when the gravitational force becomes the same order of magnitude as the electromagnetic force that operates the model circuit. This is crucial for integrity of operation of the equivalent circuit model.

The equivalent circuit model must work only by the electromagnetic force, and no other force should be involved in its operation. The size of the model circuits is practically unreachably small, and the operating time is short, to be consistent with the particle's size and the limit of resolution of time. This space-time limit, called the Planck regime, is reached when the gravitational force working among the circuit elements is the same as the electromagnetic force driving the model circuit. Beyond the minimum limit, the model circuit's operation ceases, because the circuit components cannot sustain their structure against gravity, and also their connection is affected by the gravity. Simply saying, gravity *crushes* any smaller circuit elements and their connected structures beyond the limit. The size is about 10^{-33} cm. Below this scale, there are no definite circuit structures, no components, and no connections. In the equivalent circuit model of quantum mechanics this feature must be reflected, and as a consequence, all the circuit parameter values must be probabilistic at the beginning of the universe.

The time of the circuit operation is short, but it must secure causality of the quantum phenomena. The scale of the minimum time in the Planck regime is estimated to be about 10^{-43} second, estimated by dividing the Planck length 10^{-33} cm by the present light velocity, about $3x10^{10}$ cm/sec. From the equivalent circuit model's operation, this time limit estimate has a doubt.

In this ultimate limit of the space-time, no object had *well-defined* shape, and the time length lacked its conventional meanings. Since no circuit component had definite shape and parameter values, probabilistic parameters, instead of definite parameters, characterized such idealized circuit model operation. The quantum equivalent circuit model must be built and operated subject to such an exotic condition, and these features still exist at present in the quantum phenomena of the elementary particles, whose real size is of the order of the Planck length. This is the world where *information* characterizing any physical phenomena never existed. The equivalent circuit worked not on the definite parameters, but all the parameters were *probabilistic.* It is interesting to imagine the equivalent circuit model in such an extreme limit.

Suppose that I make a circuit model of quantum phenomena in this space-time limit. The components that build the equivalent circuits are still triodes, resistors, capacitors, and inductors. The smallest elements are triode and resistor. A triode controls flow of the modeled substance, the particle and its energy. The flow is equivalent to a quantitative parameter corresponding to current in the circuit model. The controlling force of motion of the flow is the qualitative parameter like voltage. The triodes and resistors are the smallest circuit components, because they need not store the quantitative objects, the equivalent of electrons in the circuit model, in the component's structures. It is more like a small workbench than like a large warehouse. Physically, triodes and resistors were voltage-controlled small holes that either allowed or blocked passage of the particles. Resistors were modified triodes that probabilistically pass or stop the flow. The size of the hole's diameter must be comparable with the particle's size, that is, the Planck length.

I follow the universe's development, starting from this earliest stage that I call phase 1. The equivalent circuit could be built only by the smallest size components, i.e., triodes and resistors. According to my principles mentioned before, I should be able to build the model of any phenomena that occurred in the phase 1 universe by the two components only. Any equivalent circuit, such as a logic gate, in this phase worked *instantly*, that is, not exactly zero, but in the minimum unit of time to secure the input-output causality. Time's ultimate limit was reached to secure causality but not its definite length. There was no circuit element setting the time scale less than this *minimum time* that assured the model's causal relation. Time is a continuous parameter that cannot be *quantized*.

From this viewpoint, the Planck length divided by light velocity is not the minimum unit of time. The phase 1 universe was the *timeless* world. Any equivalent circuit in this regime was built and broke down *instantly*. Since such components had no definite shape and parameter values, they must have worked probabilistically. This was the model's feature of the phenomena in the earliest phase 1 universe, which still remains in the present quantum mechanical phenomena, when the quantum states are built or destroyed. Quantum transition occurs instantly. This feature has been a mystery of the quantum world, as it had been the theme of the debate between Bohr and Einstein on the interpretation of the quantum entanglement experiment. Yet that is not a mystery from the operation of the equivalent circuit model–based thinking. The signal that controls the probabilistic parameter propagated, and still it does at infinite velocity. The quantum state is a probabilistic existence; it builds up and breaks down instantly by the probability signal even at present.

Suppose that the universe became ten times larger in size. As the size of the universe increased in this phase 2 stage, the circuit components began to have better defined structure and parameter values than in the phase 1 universe. That is, the random, probabilistic operation of the phase 1 universe began to display some orderly features. The probabilistic parameters still dominated, but it began to change gradually to quasi-definite (meaning somewhat orderly) parameters. The probability was gradually changing to information during this phase.

In this phase, one other component emerged in the equivalent circuit model: a capacitor consisting of a pair of plates facing each other. Its value was given by the plate's area divided by the plate's gap. The gap was small, because the electric field lines terminated by the charges on the pair of plates, and the electrostatic force could be made strong relative to the gravitational force.

The device was tolerant to gravitation. In the dimensional analysis notation of using the length dimension symbol [L], capacitance is proportional to the dimension $[L]^2/[L] = [L]$. This element emerged when the universe's size increased such that the capacitor plate's size could be, say, ten times larger than the plate's gap. Now the triode and the capacitor set the time scale longer than the absolutely minimum time scale required to secure the causality. Behaving distinctly from space, time does not have the minimum unit; it is a mathematically continuous parameter, and the order is set by whatever the small difference.

Time as we know it now emerged in this phase 2 universe, as the ordering parameter of the triode-capacitor equivalent circuit activity. This triode-capacitor time setting of the expanding universe consumes the free energy. This was possible since the free energy density of the phase 2 universe was still quite high. Logic gates executing a *sequential process* emerged in the phase 2 universe, and the complex sequence of events was sequenced by the emerging phase 2 universe's time. This feature continues to the present. In the phase 2 universe, all the prototype structures of later physical phenomena emerged. Yet the circuit parameters were dominantly probabilistic. This is reflected in the general form of the basic equation of quantum mechanics, the Schrödinger equation, which covers all physical events, and it has the model of the triode-resistance-capacitance equivalent circuit (sections 1.05, 2.01, and 2.02).

The equivalent circuit model contains no inductance. Then the light velocity signal propagation delay time was not included in the phase 2 equivalent circuit model. *Light velocity did not exist.* This is the reason why I did not consider the time's minimum unit was the Planck length divided by light velocity. The phase 2 universe's features continue to the present; the quantum mechanics equivalent circuit model contains only triode, resistor, and capacitor. This is also the feature required to propagate the probability setting signal at infinite velocity.

Let us consider that the size of the universe expanded 100 times the Planck length. I call this the phase 3 universe. This allowed the definite operation of some circuit model, since the size of the components increased and in the end approached macroscopic size. A signal carrying definite information emerged from the probability signal. Along with information transmission, inductance, the only remaining circuit element, emerged. Inductance value is proportional to the area of the current loop, which has the dimension of $[L]^2$. Why does inductance require more space size than capacitance? Inductance requires wire, whose diameter must be enough to pass the electrons, and the magnetic field line must close itself in the open space above, below, and to the sides of the current loop. The emergence of inductance was concurrent with the emergence of information from probability.

Inductance-capacitance delay time was negligible in the earlier phase 1 and phase 2 universe. The probability signal propagated at infinite velocity. This feature is not mysterious at all, if we consider even everyday probability events like an idealized lottery operation. In the phase 3 universe, inductance is carried by any object having nonzero size. Inductance-capacitance became the deterministic time-keeping factor. This time-keeping factor is associated with any object that has larger size, and furthermore, it does not require consumption of free energy of the universe to set the time scale, as triode-capacitance time. This is the ultimate time-setting element when the universe's energy density is low, as it is now. The inductance-capacitance of the space set the velocity of light, which limits the maximum velocity of any object or information carried by it. Information can be carried only by energy, since the signal must trick the receiver to deliver information. During the phase 1 universe, when the system's operation is probabilistic, the signal needed not crick the receiver, and therefore the signal carrier needed not be energy. Such a signal carrier couldn't be observed externally. The information-carrying signal can be intercepted and is observable by a receiver.

If the velocity of light were calculated in the phase 2 universe (if such a parameter were calculated as the inverse of the square root of the product of the space's inductance and capacitance), it tended to be infinity in the limit of the beginning of the universe. It depended on the size of the universe by the relation of the dimensional expression $[L]^{-3/2}$ which diverged when [L] tended to 0. In the phase 3 universe, the delay time due to the light velocity sets the scale of the time. As such, this became a universal time-setter. The light velocity limit remained as the size of the universe increased further to the present size. We are still in the phase 3 universe.

From this viewpoint, a significant conclusion is that relativity became the basic principle of physical phenomena starting from the phase 3 universe, but not before. A simple interpretation of this feature is that there was no time or space at the very beginning of the universe. Then why was relativity relevant to deal with the situation? Quantum laws are more basic and archaic than relativity laws because they emerged earlier than the phase 3 universe.

Newtonian classical mechanics with instant action at a distance reflects the earliest phase 1 universe's state, but the parameters are definite since the size of the object is so large that probabilistic effects are negligible. The capacitance-inductance delay time set the time scale by relativity. Thus, relativity modified the Newtonian mechanics. The pre-relativity classical mechanics model is a combination of the two historically very different times of the universe's age. It reflects the earliest phase 1 universe's instant action, and the definite parameter value of the phase 3 universe, but not the time set by relativity. That is why it has the simplest mathematical structure. Relativistic classical mechanics reflects the phase 3 universe's relativistic features by removing the phase 1 universe's instant action but still adopts the definite parameters characterizing the physical state. The mathematics became more complex than Newtonian mechanics. Yet classical mechanics ignores the probabilistic features of the earlier phase 1 and 2 universe.

Nonrelativistic quantum mechanics, whose basic equation describes the character of the phase 1 and 2 universe when only the triode, resistor and capacitor existed, reflects the early universe's state at the time when capacitance had emerged but not inductance. The probabilistic nature of the earlier universe is its essential feature, including the probability signal's instant propagation. The model of nonrelativistic quantum mechanics can be built by taking this feature as its basis. This has a crucial point in the quantum effect modeling. The logic gates that model nonrelativistic quantum mechanics do not include the light velocity delay time, since inductance had not yet made entry into the equivalent circuit model. Then, even if the gate has length L, the gate should not carry light velocity delay time L/c (where c is light velocity).

The gate delay is set by the time of charging the capacitor by the triode. This is because the circuit model still had no inductance, but the triode parameter was calibrated by including the effects of inductance in the phase 3 universe. This is a crucial point in the later equivalent circuit model development in my work. Without this feature, some elementary quantum effects, such as instant collapse of the wave function and the quantum entanglement effect, cannot be explained.

In the phase 3 universe, the parameter values of the equivalent circuit model were calibrated by referring to the emerging relativistic feature of the light velocity limit, which set the present space and time measure. The triode-capacitor equivalent circuit structure of phase 2 universe is still kept as it was, but the velocity of an information-carrying signal propagating the buffer chain is limited by the calibration of the buffer parameter involved, that is, the buffer driving current (section 2.11). Any information-carrying signal cannot propagate faster than the light velocity through the buffer chain.

Relativistic quantum mechanics, the most advanced quantum theory, includes the features of the phase 3 universe and all the earlier phases' features, which exist at the present time. As the most precise description of the *present* quantum phenomena, probability effects still dominate because of the small size of elementary particles, but the inductance-capacitance (light velocity) delay time, the relativistic effect, is explicitly included in the theory. Enormous precision of the theory was reached. Yet it comes with a cost, however. Regarding the theory of cosmology of the earliest universe that tries to combine general relativity and quantum mechanics, what I mentioned above is a diametrically opposite viewpoint of the earliest universe. The general relativity–quantum mechanics combined theories seem to have all sorts of mathematical difficulties of divergence. I expect that relativity phases out to explain the earliest universe prior to phase 3, that is, before inductance emerged as the equivalent circuit component.

I go into one more modeling detail. My equivalent circuit model may raise a question: are triodes, resistors, capacitors, and inductors the only components that are required to build the equivalent circuit model of any physical phenomenon at any phase of the universe's development? Can there be anything else? My answer is no. This is because the physical parameters are one of two kinds; a volume parameter such as current and an intensity parameter such as voltage. The resistor gives their proportionality. By the capacitor-inductor pair, the differential of the one type parameter gives the other type parameter. Then the three components are enough to model any physical phenomena at any phase of the universe's history.

1.07 Quantum Laws in the Universe's History

In the last section, I stated my basic belief that any physical phenomena, classical or quantum, can be simulated by properly setting up the equivalent circuit model. To model quantum phenomena, the order of emergence of the circuit components of the model in the early history of the universe matters. Now I examine the characters of the equivalent circuit from each phase of the universe's development. This observation explains a curious feature of the quantum law: that they seem to consist of several qualitatively different laws assembled in parallel. The most archaic feature remained in the simplest modes of operation of the quantum physical phenomena: build-up and collapse of the state of the wave function, which happens instantly.

In the phase 1 universe, only triodes and resistors existed. Since they were in the Planck regime, none of them had properly defined structure and parameter values. If they built the quantum effect's model, the model's input-output relationship was purely probabilistic; if something happened at the input, something would happen at the output. The only significant feature was causality, that the input's cause was followed by the output's effect. What the cause and the effect were never mattered. Chaos dominated in this phase 1 universe.

Time is a parameter like a real number. Two times indicated by a pair of real numbers have the order of its progression, but the difference between the pair of the numbers is irrelevant in determining the order. Causality originated by this absolutely minimum ordering of time as the real number parameter. This continued to be so from the phase 1 universe to the present phase 3 universe. In simple terms, there was no *quantum* of time. Planck length divided by light velocity is not the time quantum. I have showed that the *light velocity* diverged at the beginning of the universe.

Events in the phase 1 universe had little variety; the operation of the circuit model was to build an unspecified state and then destroy it. Generally, the destruction of some structure, including macroscopic structures, can happen instantly. Some breakdown needs no propagation time of a *crack* from one end of a body to the other. In the burning of a fuse, if the temperature is raised uniformly, at a certain point it burns all at once. This feature is observed in quantum mechanical phenomena as instant collapse of quantum states, which can also be interpreted as the probability signal's infinite velocity propagation. The same is true in the buildup of a quantum state. A quantum state is a probabilistic existence that is built and destroyed instantly. The structure becomes observable only if the particle rides on the state. In the phase 1 universe, random and instantaneous buildup and destruction of arbitrary states is the only mode of operation. This phase 1 universe's probabilistic feature remains to exist at present, when the wave function collapses.

The phase 2 universe was not homogeneous, but the available space was increasing to accommodate complex structures. The buffer chain of section 1.05 became the model of the particle propagation path. The space measure was the length of the buffer, and the time measure was the buffer's switching time. Yet each buffer and the load capacitor still had no definite parameter characterizing them. The parameters were different from buffer to buffer, since the quantum space was still highly nonuniform. There was no universal space-time measure to specify the circuit model operation. Because of that, probabilistic development of the circuit operation dominated, but it became more orderly than in the phase 1 universe. That is because the capacitor created two effects. The circuit output was not instantly responding to the input change any more, since it took time to charge/discharge the capacitor by the triode. It became possible to distinguish the effect from the cause clearly. Then some order emerged among the composite state's structure and operation. The triode-capacitor time delay, the *local time*, emerged as the parameter to relate the causality of the *related* event's sequence. This time is *sensed* by the particle, but not by the external observer.

In the phase 2 universe, the prototype of the two basic parameters, the local time, that is the switching time of a buffer, and the local spatial coordinate, the size of a buffer, emerged. The triode-capacitor time was the *local time* that arranged only the input-output sequence of the related sequential events. The local time depended on the individual buffer's triode-capacitor time constant.

To develop complex quantum phenomena, the elementary particle's states must be integrated. As a physical process, this was the generalized diffusion process as I discussed in sections 1.03 and 1.04. The development of the quantum world took the similar form as the classical diffusion process, since any diffusion process has the basic feature; that the gross order emerges from the components' disorder. As the phase 2 universe expands, semi-predictable state sequences emerged, and that was governed by more definite operation of the composite objects. The basic theoretical framework of the quantum mechanics was established in the phase 2 universe. That is why the Schrödinger equation retains mathematical similarity to the diffusion equation, modeled only by resistive and capacitive circuit components.

The operation of the model of the particle during the phase 2 universe could not be described by the universal space and time measures, since each object, the buffer, still did not have precisely determined parameters. The time was its own *local time*, only to order the particular event's sequence not shared by all the phenomena of the quantum world. This means, each particle *senses* its own space and time. Here arises a basic question: Should the state of the early universe be considered by relying on the time-space concept that is at the present time? Or on the space-time as sensed by the particle itself that is going through the state change? I believe the latter is the case.

This feature shows up most clearly in the example of section 3.05. There, the parameters, the *length of a buffer* and the *velocity of flight* of a particle in the double slit experiment, are valid only for each particle in its own separate paths. They are parameters sensed by the particle, and not by the external physicists. The particle carries its own sense of local time and of local space in its path. The reason for the particle propagation model's failure is that the two particle paths' parameters are implicitly assumed to be the same by observing physicists. Actually they must be considered as the distance of the path and the velocity sensed by the particle itself. A pair of buffer chains, each having a different number of buffers, could have same total length and the same traveling time sensed by the particle. If this is not admitted, the particle model's failure is naturally expected. The mechanism that explains this strange feature is the uncertainty principle.

The uncertainty principle prescribes the somewhat more *ordered* feature of the phase 2 universe's phenomena than the chaos in the phase 1 universe, yet the probabilistic feature of the individual particle's path clearly dominated and displayed the space-time (or momentum) uncertainty. A particle's position and its velocity cannot be determined precisely because of the local time and local spatial measure. The time and space measure uncertainty is the basic cause of the failure of the particle model to explain the quantum phenomena and the adoption of the wave model instead in quantum mechanics.

Since the phase 2 equivalent circuit model emerged and established itself before the emergence of inductance, the circuit model operates on the time scale determined by the triode-capacitance delay time. The triodes and load capacitance were both not uniform over the particle's propagation path. The probabilistic features associated with the phase 2 universe were not conforming to the relativistic constraint of locality. Because of that, the component building the path, the buffer's delay time used to model the phase 2 quantum effects did not include the propagation delay time of the signal going through the buffer's spatial length by the light velocity, thereby allowing infinite velocity of probability setting signal propagation if the buffer's load capacitor is zero.

This is a permanent feature of the basic quantum phenomena that emerged in the phase 1 and 2 universe. An example is this: if several cascaded buffers are not capacitor-loaded, they can be compacted to a single buffer, the buffer chain length decreases, yet the circuit works exactly the same way as before compacting. The effective velocity of the particle increases by compacting. This gives us an impression that the spatial topology is affected. We see this curious feature emerging from the uncertainty principle, based on the probabilistic system operation. Apparent spatial and time *distortion* is due to nonuniformity of the local space and time measure of the phase 2 universe. The effect is a remnant of the earlier phase 1 universe. In the phase 2 universe, the model's logic circuit structure, and not its physical size and the time length, determines the phenomena.

The other effect of the phase 2 universe was that the prototype of all the complex dynamic effects emerged. The quantum *prototype* of all the presently existing physical phenomena emerged during this phase 2. That is the reason why quantum mechanics covers everything existing in the physical world. The circuit model that includes triodes and capacitors in nonrelativistic quantum mechanics operates on the probabilistic and nonrelativistic modes. Signal transmission creating physical phenomena is explained by the model including only of triodes and capacitors. This feature changed by including inductance in the following phase 3 universe. The phase 2 quantum space could not provide a universal time scale because capacitance is a localized component; any capacitance value can be created by adjusting its plate's gap. Inductance is different; it requires a wide enough spatial area to accommodate the current loop. To make any inductance value, we do need a uniform and wide space. This requires the phase 3 universe, which provides a wide and uniform space.

The large circuit component, inductance, emerged in the phase 3 universe. The light wave, that propagates indefinitely along the free space's inductance-capacitance transmission line, controlled the quantum space activity and set its global time scale. The quantum system became orderly. An information-carrying signal emerged.

Now the space has two different kinds of signals. *Information* of the system is carried only as fast as the light velocity, but the previously existing probabilistic signals are still carried at infinite velocity. Light velocity became the established measure of space and time in the phenomena of physics. This feature is reflected in the equivalent circuit model in two ways. First, since global uniform space and time measures were established, the randomness features of the circuit model structure and parameters of phase 1 and 2 universe were reduced. Since there is a universal measure of space and time and uniform large space, all the buffers of a uniform signal path must have the same parameter values, so that they can be cascaded to make a uniform particle propagation path through the expanded space.

If inductance is included, the light velocity c is given by the $c = (\varepsilon\mu)^{-1/2}$, where ε and μ are the electrical and magnetic susceptibility of the phase 3 universe's uniform and free space. Then the velocity of the particle v is given by its fraction, $v = \gamma c$, where γ is a factor less than unity. The particle velocity must be referred to the universal constant, light velocity, by the observer. This means that velocity calibration by light velocity is in effect, disregarding the velocity sensed by the particle itself by the observing physicists. This creates confusion between the particle and wave models of quantum mechanics. The wave model, instead of the particle model, took over quantum mechanics.

The emergence of inductance had yet another effect, reflecting the relativity features. Energy or mass cannot move faster than light velocity. This means not only that the parameters of the buffers are made uniform in the uniform phase 3 universe's space, but that all the parameters must have properly set limits. This means calibration of the particle path's parameter, that is, of the buffer's parameters to meet the light velocity limit requirement, as I discuss in section 2.11. By going through the steps, the emergence of inductance makes the transition from the quantum realm to the classical physics of macroscopic objects. The reduced probabilistic features, universal time and uniform space, created the quantum physics of the phase 3 universe. A large object's motion can be described by definite parameters instead of probability and by the universal time in which the basic laws of classical physics apply.

The earlier phase 1 and 2 universe's theory of quantum phenomena operates in the phase 3 universe by using the *calibrated* time, space, and model circuit component parameters of the phase 3 universe. The significant change of quantum mechanics due to the emergence of inductance was that, by using the calibrated parameter values, the signal carrying information cannot propagate faster than light velocity. The inductance effect altered the quantitative feature of the model by including universal space and time measures, but still, the basic probabilistic features remained intact. The basic logic circuit structure of the quantum effect model consisting of triode and capacitance remained the same. Yet the circuit parameter values are calibrated by referring to light velocity. The quantum mechanics' qualitative structure (resembling the equivalent circuit model), probabilistic operation, and instant probability signal propagation remain unchanged even in the phase 3 universe by making the theoretical description relativistic. Thus, three different phases of the universe's development provided what appears to be three independent principles of quantum mechanics working in parallel. As Einstein felt, the quantum theory is not clean; it gives the impression that something is missing and therefore it appears like the assemblage of several independent principles.

Quantum mechanics started from the original chaos, but it is a precision science now, because it uses time and space measures that have been set in the phase 3 universe, while keeping the theoretical framework of the phase 1 and 2 universe. The quantum mechanics of phase 2 was not precise in this sense, but it builds up the skeletal theoretical structure, including the probabilistic features. Including inductance in the equivalent circuit model not only conforms the theory to relativity but makes it precisely predictable. The present front of quantum physics adopts this integrated theoretical structure.

The enormous precision achieved by this integration has positive and negative effects. The first is that the theory describes the *present* state of quantum object with extreme precision. Its cost is that integrating general relativity and quantum mechanics in the attempt to explain *the state of the early universe* cannot be achieved yet, and it is still unknown whether that is possible. There are all sorts of mathematical difficulties and strange problems emerging in the theory, and there is yet no generally accepted integration since 1970s. According to my historical model-based thinking, relativistic features could possibly phase out as the time approaches the universe's beginning, that is, at the time of changeover from phase 2 to phase 3 universe development, when inductance was included in the model.

Digression: Equivalent Circuit Model

In the equivalent circuit model used in classical mechanics, circuit size does not matter. There may be a question why the equivalent circuit size must be scaled down to model quantum phenomena. The reason is that in the limit of small scale, both time and space are shrunk to the limit, and that feature must be reflected in the model. Relativity assumes extended space and unlimited time. This assumption cannot be carried there. If both parameters tend to zero, it is necessary to reflect this feature in the model.

It might be thought that the active triode is not the smallest component. This is because of the lingering historical bias of the oldest triode, vacuum tubes. Most of us have never seen how small the CMOS triodes are in the VLSI processor. The active part of the CMOS triode is more than 50 percent of its volume. Comparing to that, a vacuum tube's active part is less than 5 percent of its volume, and the rest is vacuum, glass bottle, and terminal structures. Then the triode appears naturally larger than capacitors and some inductors. This is a wrong impression of the nature of triodes.

1.08 Substance and Character of Elementary Particles

The natural way to classify the parameters of an elementary particle is to distinguish its *substance*, its *characters*, and its dynamic parameters. The substance of a particle is the particle's energy and mass, both of which are forms of energy according to relativity. A particle carrying substance cannot move faster than the velocity of light. There is the mechanism of exchanging kinetic and mass energy carried by an accelerating or decelerating particle.

The particle exists in the potential field, and it gives or takes energy to and from the potential field. The particle's motion in the field is trading the energy between them. The rate of exchange between the potential energy variation and the particle's energy variation is set by the particle's character charge. The charge's varieties are the spin and the electrical, weak, and color charges. The character's charges have magnitude, polarity, and direction of spin. Any character's charge determines the type of the force exerted to the particle. The exerted force changes the particle's energy, but the charge remains unchanged. The substance and the charge are the two distinctly different characters of the particle, and they have different origins and different way to responding to the environment, in affecting the particle's dynamic state.

Elementary particles carry dynamic parameters like position, velocity, energy, and momentum. The particles are so small that not all of their dynamic parameters are precisely measurable by whatever means. That is the reason why probabilistic characterization is required to study the particle's behavior. There are complementary pairs of dynamic parameters; accuracy of measurement of the one affects the accuracy of the other. The uncertainty relation between the particle's position x and its associated momentum p_x is the best known. Their measurement accuracy δx and δp_x satisfy Heisenberg's uncertainty relation $\delta x\ \delta p \geq \hbar/2$, where $\hbar = 1.054 \times 10^{-34}$ is the joule-second is the universal constant that emerges everywhere in quantum theory. This relation can be derived by the thought experiment of illuminating the particle by light. If light's wavelength λ is made short to determine the precise particle location, then the photon's momentum $2\pi\hbar/\lambda$ is large, and that affects the particle's momentum. Energy and its measurement time, and angular momentum and the angle are related by the similar complementary relations.

There are parameters that can be measured definitely without uncertainty. They are the charges of the characters. The magnitude of an electron's angular momentum, spin, and electrical charge are precisely measurable. A remarkable feature of such parameters is that they are all *confined* within the particle's small size. The size is considered as a sizeless point, but realistically believed to be the Planck length of about 10^{-33} cm. An illustrative example of the confined and not confined parameters is the particle's spin. Spin is the angular momentum of the particle itself, and its magnitude is *confined* within the particle and is definite. It is an integral multiple of $\hbar/2$ for fermions. Since angular momentum is a vector, the line specifying the direction can stretch out from the particle to the space outside the particle. The angle specifying the line is not confined within the particle. Then the spin direction is probabilistically specified. By applying magnetic force, the direction of the spin of any electrically charged particle can be changed. Spin is different from electronic or any other charge.

A peculiar feature of the confined parameters is that they come as a pair of positive and negative values, and if the pair combines, the parameter vanishes. This is the lowest energy state. This feature suggests that the character's charge was created by splitting the neutral mother particle into a pair of polarized daughter particles (section 3.10). The splitting may involve sorting the charges and separating them into the split pair. This suggests that there are equal numbers of positively and negatively charged particles. Splitting the particle's substance may also create the gravitational force. In this case, there is no sorting. There is no negative mass, and the force is always attractive. The force always tries to reduce the object's surface area. This explains why gravitationally coagulated objects always become spherical in shape, except when the force is too weak to overcome the material's natural stiffness or centrifugal force.

I need a caveat. Any dynamic parameters are not confined within the *point* particle. Energy, the particle's substance, appears to be a dynamic character, but it seems to be confined within the particle. Yet it is influenced by the force acting on the particle. The particle's substance is continuously interacting with the potential field with the charge as the media, so energy is effectively spreading out to the outside of the particle. As such, it is not a definitely measurable parameter.

1.09 Limit of Splitting Material

Any complex information can be split down to a composite of *1* bit. By *1* bit, a single selection of object or state A or B can be securely executed. If *1* bit is further split conceptually, then probability emerges (section 1.01). Corresponding to that, how much can any material be split? What is the equivalent of the minimum unit of *1* bit of any material? Here is a historical question; Greek philosophers thought that material's *substance* can be split to atoms, but the substance depends on the energy given to or taken from it, and energy is an externally controllable parameter. Then the minimum bunch of energy or the mass cannot be identified as the atom in the Greek philosopher's sense.

Let us review this point. Substance is some form of energy. Mass is the most popular form of the substance. The character's charge is the attribute of the substance that specifies the force acting on the substance. Force acts on the substance carrying the charge, and as a consequence the substance is changed by the force. This feature makes it impossible to consider the minimum substance as the basic unit of material. The mass of an electron is not a basic physical parameter. It increases by acceleration.

Charge is a quantitative parameter in composite or macroscopic material. A macroscopic object carries a lot of charge. When the material is split, the charge is also split and, at the limit of splitting, becomes the charge of the elementary particle. The elementary particle's charge cannot be split any further, so it becomes the indicator specifying the elementary particle. This is an interesting case of a macroscopic quantitative parameter change to a microscopic qualitative parameter by splitting. In more secular terms, splitting material to elementary particle is not like cutting meat into small pieces. It is rather restructuring a building with multiple doors into town houses, each of which has only one door. Each door allows entry or exit of persons into or out of the town house. The number of persons in the house is the equivalent of the substance.

The minimum (undividable) parameter carried by the elementary particle is the unit of character's charge, such as spin and electrical, weak, and color charges. They carry definite value, because they are the identifier of the particle to be elementary. The charges work as the media of exercising the force but do not themselves change by the applied force. Then any material can be split to the unit of the character's charge, which can be expressed by *1* bit of information by using their physical value as the unit. Thus, a quark cannot be isolated, because the electrical charge is fractional. The minimum charge's physical value (actual number) is measured by whatever physical units, such as CGS or MKS unit. Material can be split to the limit that each particle carries the unit charge of its character, specified by *1* bit.

The substance associated with the unit charge of various particles in a static state cannot be specified by the type of the charge. Unit electrical charge is carried by electrons, muons, and tauons. They make the lepton family, and their masses are 0.511MeV, 105.7MeV, and 1777MeV, respectively, almost covering a range of 3500 times. The same spin angular momentum $\hbar/2$ is carried by the heaviest lepton and the lightest neutrino. A neutrino's mass is not zero but is almost unmeasurably smaller than that of an electron. The character charge and its substance are independent attributes of elementary particles. This is the point I discuss later in detail in section 3.08 by taking the case of spin. Any material can be split into elementary particles until the character's charge carried by each particle, measured by any unit system, is specified by one bit of information, such as *1*, *0*, or *-1*. In this sense, both material and information have similar limits of division. The only difference is that an information unit can be split further conceptually into probability, but there is no fragment of a particle's charge observed experimentally.

Leptons carry unit spin and unit electric charge. If splitting of macroscopic material is executed arbitrarily, it is not likely that the limits of different charges are reached simultaneously. Yet why is the limit reached simultaneously? In reality, there are certain relations between the amounts of different unit charges. As for the electronic charge and spin, there is a relation $e^2/\hbar c = 1/137$ This dimension-less number, 137, has never been theoretically explained. Since e is the unit electron charge and $\hbar/2$ is the unit spin, there is a relation

$$\textit{unit electron charge} = 0.1208(\textit{light velocity x unit spin})^{1/2}$$

and the splitting of the material is executed *simultaneously* until both limits are reached simultaneously. That is the reason why all the leptons have unit electric charge and unit spin. Yet this relation is for leptons only. For neutrinos, the factor *0.1208* is replaced by *0.0000*. The relation among the unit charges has not yet been explained theoretically, but there are similar relations for different charge types.

2

EQUIVALENT CIRCUIT MODEL OF QUANTUM PHENOMENA

2.01 Basic Linear Equivalent Circuits

Quantum phenomena's equivalent circuit model is a combination of linear and digital circuits. The linear circuit model is to display the quantum wave features, supporting the principle of probabilistic superposition of quantum states. The digital circuit model highlights the classical particle motion of the wave packet created by superposition of the waves and reveals some hidden nature of quantum particles, fermions and bosons (section 2.09).

The Schrödinger wave equation is the first rank in time derivative and the second rank in spatial derivative. It has the same form as the diffusion equation. Its one-dimensional form of the particle moving in no external field is given by

$$\hbar \frac{\partial \psi}{\partial t} = - \frac{\hbar^2}{2m} \frac{\partial^2 \psi}{\partial x^2} \text{ or } \frac{\partial \psi}{\partial t} = iD \frac{\partial^2 \psi}{\partial x^2} \quad (1)$$

where $D = \hbar/2m$ and where ψ is the wave function, whose amplitude squared gives the probability to find the particle moving along a straight line at location x at time t.

As the equivalent circuit model of this equation, I consider a conventional distributed parameter resistor-capacitor ladder circuit of Figure 2.01.1(a). As discussed in section 1.06, it consists of resistor and capacitor only and includes no inductor. Let the voltage at node x at time t be $\psi(x,t)$. Its increment $\delta\psi$ during time δt is given by the equation

$$(C\delta x)\delta\psi(x,t)=\delta t\{\psi(x+\delta x,t)+\psi(x-\delta x,t)-2\psi(x,t)\}/(R\delta x)$$

where the parameters C and R are capacitance and resistance per unit length of the RC ladder circuit, and {...} equals $(\partial^2\psi/\partial x^2)(\delta x)^2$, respectively. Then I obtain the conventional diffusion equation

$$C\frac{\partial\psi}{\partial t}=\frac{1}{R}\frac{\partial^2\psi}{\partial x^2} \text{ or } \frac{\partial\psi}{\partial t}=D\frac{\partial^2\psi}{\partial x^2} \quad (2)$$

where $D = 1/CR$. If I compare equations (1) and (2), D of Eq. (2) is related to that of Eq. (1) by a factor i, the unit of the imaginary number.

The requirement of this model to represent a uniformly propagating particle in a straight line is that there is no increase or decrease of the amplitude of the wave function ψ.

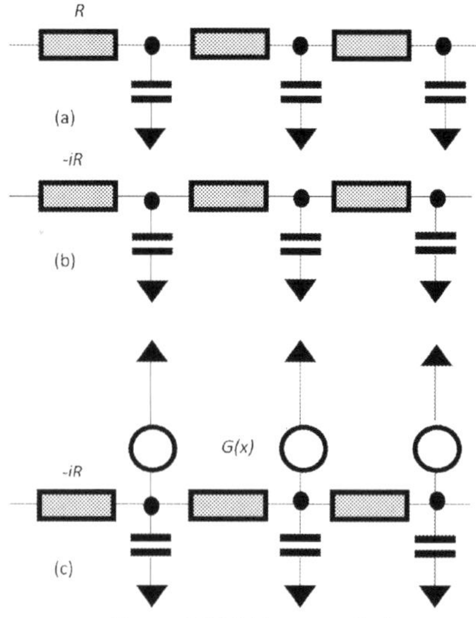

Figure 2.01.1 Linear equivalent circuit model of quantum equation

To satisfy this requirement, the factor R of the diffusion coefficient D of eq. (2) must be purely imaginary. The capacitor C must be real, because it does not consume or generate energy. Resistor R consumes energy if it is real and positive. The wave amplitude decreases. If it is real and negative, it generates energy and the wave amplitude increases indefinitely. Both cases are unacceptable. Then R must be replaced by the purely imaginary value $-iR$ to make eq. (2) match Eq. (1). The equivalent circuit of eq. (1) is shown in Figure 2.01.1(b). There may be a question whether the imaginary resistor models a diffusion effect. It does, because the complex number wave amplitude of the resistor-capacitor ladder is affected in the same way even if the resistor is real or imaginary. As a generalization of the concept of diffusion effect, the imaginary diffusion coefficient is allowed as realistic.

What is an imaginary resistor? If voltage V is applied to resistor $-iR$, current iV/R flows. The power consumed by the resistor is $\mathrm{Re}(iV^2R) = 0$, because R and V are both real numbers. An imaginary resistor never consumes or generates energy, so the wave's amplitude stays unchanged. However, the wave's amplitude ψ becomes a complex number. This ψ gives probability, that is not a directly measurable parameter. The complex wave amplitude squared is the probability to find the particle at location x at time t.

The equivalent circuit model includes practically not realizable resistor $-iR$. The equivalent circuit model covers wide range of possibilities to include such not realizable circuit components to model any physical phenomena. Equivalent circuit model represents connected structure of all kinds, in which the current flows and the voltage develops by the Kirchhoff's law. This flexibility has not been fully recognized, but is useful for modeling anything, especially quantum phenomena.

How is the potential energy that affects the particle included in the equivalent circuit model? The equivalent circuit is shown in Fig.2.01.1(c). In his figure, the resistor-capacitor ladder keeps the same structure as that of Fig.2.01.1(b), but the potential energy is included at each node by the conductance $G(x)$ from the node to the common ground. This is connected in parallel to the capacitor. The conductance is also pure imaginary number to maintain the amplitude of the quantum wave. Imaginary conductance cannot be made, but the circuit theory allows such parameter to be included in the equivalent circuit model. The conductance per unit length $G(x)$ is related to the potential $V(x)$ of the quantum system by the relation $G(x) = iCV(x)/\hbar$. Then, Schrödinger's equation

$$i\hbar\frac{\partial\psi}{\partial t} = -\frac{\hbar^2}{2m}\frac{\partial^2\psi}{\partial x^2} + V(x)\psi$$

has the equivalent circuit model shown in Figure 2.01.1(c), where $G(x)$ is the parameter that directly affects the capacitor's charge, and therefore it is indeed the measure of the potential energy.

2.02 Solution of the Schrödinger Equation

The Schrödinger equation is structurally the same as the diffusion equation, but the diffusion coefficient is purely imaginary. Its simplest one-dimensional form including no potential energy was derived in the last section as

$$\frac{\partial\psi}{\partial t} = iD\frac{\partial^2\psi}{\partial x^2} \text{ where } D = \hbar/2m \quad (1)$$

where ψ is the wave function of the particle moving on the straight line with no force exerted on the particle. I seek for progressive wave-like solutions of this equation. By separating ψ into a product of time t and location x's functions such as $\psi = f(t)g(x)$ and by substituting it into eq. (1), I obtain

$$\frac{f'(t)}{f(t)} = iD\frac{g''(x)}{g(x)} = -i\omega \quad (2)$$

The solution of this equation has the form

$$\psi(x,t)=f(t)g(x)=\exp(-i\omega t)\,[\,A\exp(ikx)+B\text{ext}(-ikx)]\quad(3)$$

where $\omega=(\hbar/2m)\ k^2$ or $k=\sqrt{(2m\omega/\hbar)}$ and A and B are arbitrary constants. The solutions are written in the propagating wave form as

$$\psi(x,t)=A\ \exp[\ i(kx-\omega t)]\ \text{or} =B\exp[-i(kx+\omega t)]\ (4)$$

where the first solution is a wave propagating in the $+x$ direction and the second solution in the $-x$ direction. They are the traveling waves whose amplitudes remain unchanged. This is the quantum wave of the free particle.

Since ω depends on k^2, the wave has dispersion. The particle composed of dispersive waves moves by the velocity of the wave packet, consisting of many such waves superposed to create a peak. Its velocity is called the group velocity, given by $v_g = d\omega/dk$. This formula is derived by noting that the maximum amplitude of the wave of the form of Eq. (4) occurs at $\omega t - kx = 0$. By differentiating this equation by k, I get $(d\omega/dk) = x/t$, and x/t is obviously the velocity of the maximum location of the wave. This is the velocity of the wave packet. I introduced the wave number $k = 2\pi/\lambda$, where λ is the wavelength. Then I obtain $v_g = (\hbar/m)k$ in Eq. (3). Since v_g is the velocity of the particle defined in the classical mechanics, I get the relation

$$\lambda = h/(mv_g) = h/p\ (5)$$

where $h = 2\pi\hbar$, and p is the classical momentum of the particle. This is the basic equation introduced by de Broglie. The Schrödinger equation is the generalized de Broglie wave equation, which governs all the quantum phenomena.

Equation (1) has another solution. If I substitute $i\omega$ iinstead of $-i\omega$ in Eq. (2), I get a solution of the form

$$\Psi(x,t) = \exp(i\omega t)\ [A\ \exp(kx) + B\ \exp(-kx)]\ (6)$$

where the exponent of coordinate x is a real number. The real number quantity within the bracket [...] oscillates by the angular frequency ω. This means that the traveling particle reached its either left or right end of the RC ladder circuit and is stuck at there. The RC ladder is symmetrical with respect to left and right.

In Eq. (1), ψ is a wave function that gives the probability of the particle's existence. The wave function ψ satisfies the equation that is mathematically the same as the diffusion equation. This similarity has an archaic historical origin in the quantum world, originating from the early point-like universe, and still retains its features to the present. The probability is $|\psi|^2$, and ψ itself is a complex number. When an observation is made, the probability becomes the real number, and the complex number vanishes from the theory. When external access is made to the probability field, the diffusion coefficient of the probability segments makes an instantaneous transition to minus infinity, the diffusion flow of the probability segment's not infinitesimally short distance sets in, and the real particle emerges from the complex number probability field, as I showed in section 1.04. $|\psi|^2$ does not satisfy the diffusion equation, but this is the generalized diffusion process.

2.03 CMOS-Based Quantum Equivalent Circuits

The linear circuit model describes the probabilistic state of quantum phenomena, described by the de Broglie waves. As the wave packet builds up the particle's image, its motion is described by a digital equivalent circuit model. Digital buffer chains and logic gates model the quantum particle's dynamics. The simplest equivalent circuit is the chain of cascaded buffers, built from two-stage CMOS inverters as shown in Figure 2.03.1(a) by the logic symbols and in Figure 2.03.1(b) by the CMOS circuit diagram. The buffer output node can be loaded by a capacitor, whose energy represents the particle's substance, the mass or energy. I use the CMOS logic circuit as the model, since they have a number of advantages to simulate quantum phenomena. The CMOS inverter is built from the idealized MOS (metal-oxide-semiconductor) FET (field-effect transistor), abbreviated as MOSFET.

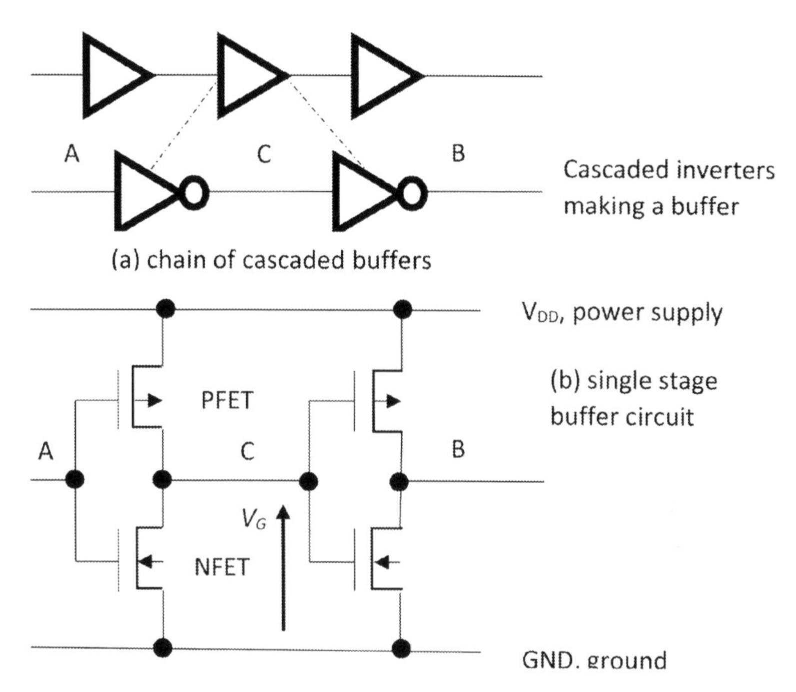

Figure 2.03.1 CMOS buffer and circuit

MOSFET comes in p-type (PFET), whose current carriers are positive holes, and n-type (NFET), whose current carriers are electrons. CMOS technology uses both FET types, and the circuit is symmetrical for pull-up and pull-down of the buffer's output nodes with respect to the power supply voltage levels. The buffers built from both FET types are cascaded to make a chain circuit, which makes the particle's propagation path, as shown in Figure 2.03.1(a) and Figure 2.03.2(a). Gate voltage V_G is defined in Figure 2.03.1(b).

The MOSFET characteristics are idealized to model the quantum effects, as shown in Figure 2.03.2(b) and (c). An NFET does not conduct unless the gate voltage V_G equals V_{DD}, that is the power supply voltage, and then it draws constant current I_0 from the load capacitor to the ground (GND). The NFET's current-voltage characteristic is shown in Figure 2.03.2(b). A PFET carries no current unless its gate voltage is zero (GND). If V_G equals zero, it carries current I_0 that transfers charge from the V_{DD} power bus to the load capacitor. The characteristic is shown in Figure 2.03.2(c). The PFET and NFET characteristics are symmetrical with respect to the V_{DD} and the GND bus.

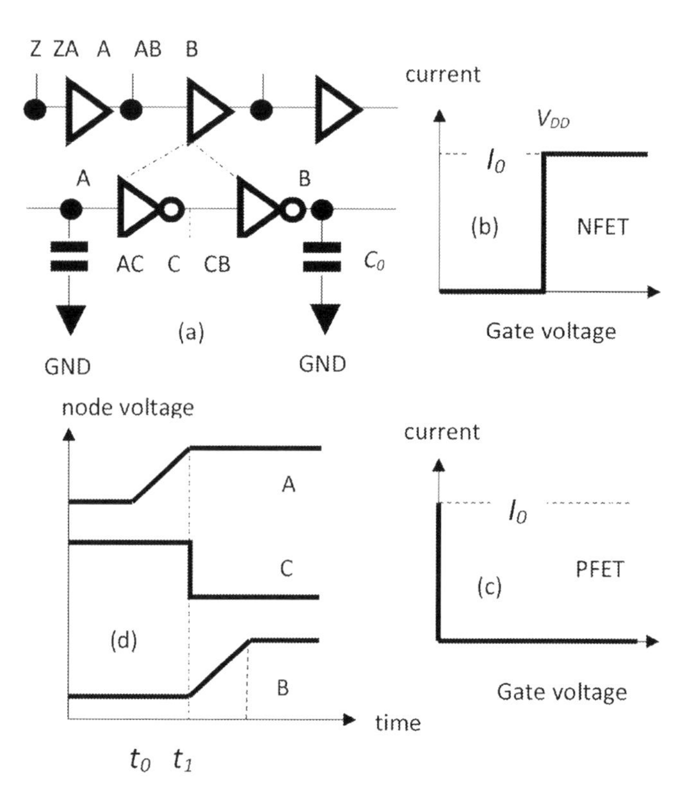

Figure 2.03.2 Buffer switching and FET characteristics

PFET's current I_0 pulls up the buffer's output node voltage proportionally to time, by charging the load capacitor C_0. Figure 2.03.2(d) shows the buffer's switching waveforms. The horizontal axis of the graph is time, and the vertical axis is the buffer output node voltage referred to the ground (GND). The buffer output node *can be* loaded by capacitor, and responds to the step function input signal driving the initial node Z. I show "can be" in italics, since the capacitor loading has certain peculiar features to model the quantum effects by CMOS circuit realistically (section 2.09 and 10).

In Figure 2.03.2(a), nodes Z, A, and B are connected by buffer ZA and AB, respectively. If node Z voltage makes a LOW (0 volt or GND) to HIGH (V_{DD} volt) stepwise logic level transition, node A voltage of buffer ZA goes up linearly with time and reaches the power supply voltage V_{DD} at time t_0, i.e., the capacitor charge-up time. This signal drives buffer AB. Buffer AB's first-stage inverter AC is not capacitively loaded. Then node C makes an instantaneous transition from logic HIGH level (V_{DD}) to logic LOW (GND) level. Then node B voltage begins to go up linearly with time by the capacitor charge up, reaching V_{DD} at time t_1. The difference of the waveforms of node A or B and of node C is due to the capacitor loading of node A and B but not C. The two inverters AC and CB of buffer AB operate differently in the switching: one instantly and the other gradually. This is a key requirement of the model allowing it to propagate the quantum particle's model waveform unchanged. By this arrangement, even a narrow, isolated pulse representing a boson (section 2.04) can propagate the buffer chain indefinitely (*Theory of CMOS Digital Circuits and Circuit Failures*, Princeton University Press, 1992). Since load capacitors at node A and B are the same, node A and node B voltage waveforms are exactly the same, if the time is shifted. The time difference $t_1 - t_0$ is the delay time of the signal going through a single buffer stage AB.

I must point out an important issue of the equivalent circuit model of any quantum phenomena. Only those capacitively loaded nodes of the circuit model like node A and B are *observable*. The capacitor carries energy, and that makes the node voltage measurable, since the voltmeter requires energy from the node to show voltage indication. Since node C has no capacitive load, it has no energy, and it is not observable. If such a node is observed, the voltmeter forces the logic LOW level (GND) to the node, and the model of the quantum phenomena does not make sense anymore. This feature emerges from the voltage measurement mechanism as follows.

An ideal voltmeter that does not carry any standby current and responds instantly to the node voltage change is a capacitor having the arrangement of measuring the force working between the two plates. This is the theoretically ideal voltmeter. Such a voltmeter was used in early electrostatic research, but practically not used anymore, because of poor sensitivity and mechanical awkwardness. Yet for basic circuit theoretical research, such a voltmeter must be assumed (*High-Speed Digital Circuits*, Addison-Wesley, 1996). Quantum effect modeling is one such case.

Even such a voltmeter has a basic problem. The voltmeter has nonzero capacitance C_V. If the node of the circuit to be measured has capacitance C_N, the node voltage drops by a factor $C_N/(C_N + C_V)$ when the voltmeter is connected. Then if $C_N = 0$, the voltmeter always measures 0 volt. This voltage is *forced* by the voltmeter to the circuit. This means that such a node is not accessible for measurement. This may look awkward, but interestingly, this is an advantage to quantum effect modeling. In the quantum phenomena circuit model, there are nodes that have no capacitance because it is not accessible for a physical reason. The voltage measurability and the quantum information accessibility become compatible. In the quantum world, not all the information of the object is available. This basic feature shows up when the equivalent circuit model is built, as shown later in sections 2.09 and 2.10.

Figure 2.03.1 shows the basic structure of CMOS inverter and buffer circuits. The CMOS logic system has varieties of its gate structure to cover the presently most complete logic system of modern digital electronics. The basic logic gates are NAND and NOR gates. Their circuit diagrams are in Fig.2.03.3 (a) and (b), respectively. By them, a complex gate such as an exclusive OR gate can be built as EXOR(A,B) = INV[NOR{NAND(A,INV(B)),NAND(INV(A),B)}]. An advantage of a CMOS circuit is that it carries, ideally, no standby current in the static state. The power is consumed only when the circuit is exercised by switching. This feature is required in quantum effect modeling.

The advantage of the CMOS in building the model of quantum mechanical phenomena is that the circuit is symmetrical with respect to the pair of power supplies. Then, particle and antiparticle can be shown naturally. CMOS also allows an electrically floating node to represent an uncertain state.

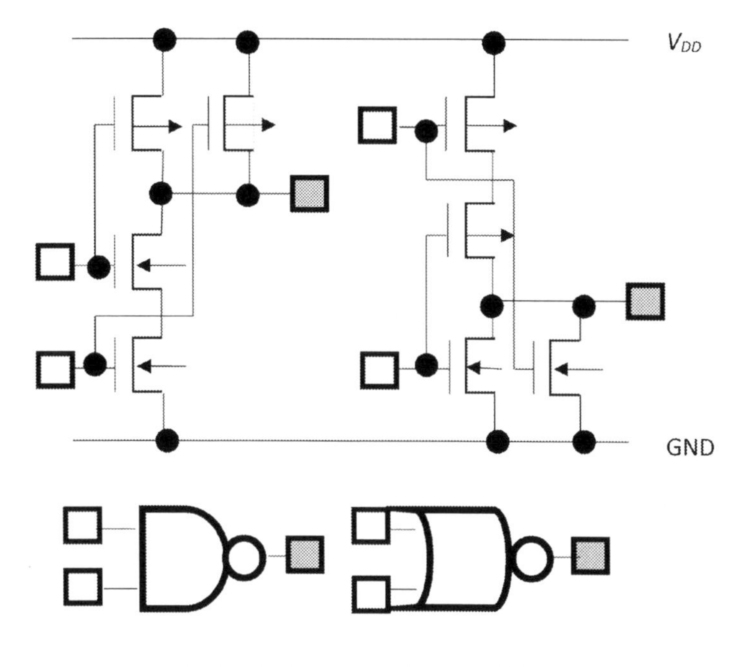

(a) CMOS NAND gate (b) CMOS NOR gate

Figure 2.03.3 Basic CMOS logic gates

What I describe in this section is the simplest explanation of a CMOS-based model of the quantum effects. The model must be revised in certain details, specifically in the operation of the load capacitor of the buffer that carries the substance of the particle, as shown later in sections 2.09 and 2.10.

2.04 Digital Waveforms and Particles

What are the basic digital waveforms that propagate the buffer chain? There are two waveform types: step function and isolated pulse. Each type has two variations in the polarity of its spatial profile, shown in Figure 2.04.1. The signal sources are on the left side (not shown), and the waves travel to the right. Any digital wave can be composed from the four basic waveforms. The pair of dotted horizontal lines in the figure, the HIGH and LOW voltage levels, holds the states of the particle. The HIGH and LOW voltage levels are called the particle and the antiparticle world, respectively. How do the step function and the isolated pulse represent the quantum particles? There are two types of elementary particles, fermion and boson. Each type has a pair, particle and antiparticle. In a single digital state, only one step function can fit in, but many isolated pulses can fit in. In a single quantum state, only one fermion can fit in, but many bosons can fit in. From this feature, the step function represents a fermion, and the isolated pulse represents a boson.

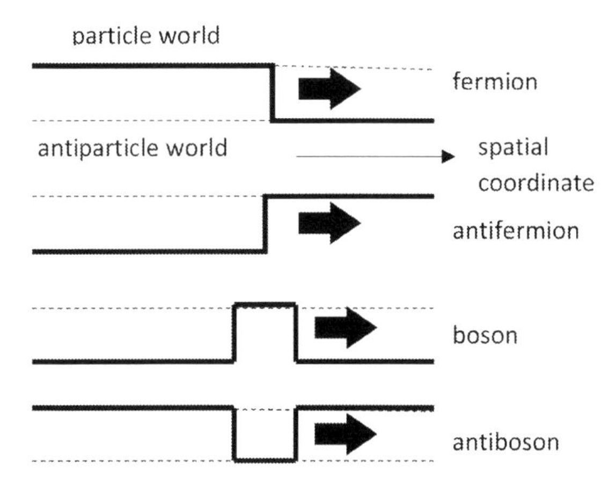

Figure 2.04.1 Digital waveforms of fermion and boson

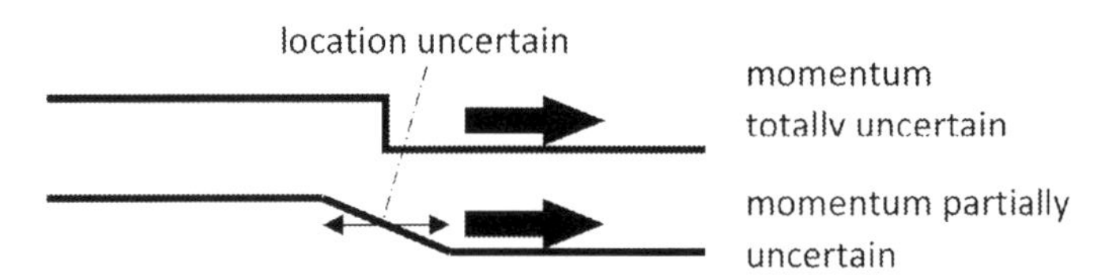

Figure 2.04.2 Fermion location uncertainty

The particle moves to the right. By the uncertainty principle, if the particle's position is definite, the particle's momentum is completely uncertain, as shown in Figure 2.04.2 (above). In the figure (below), the particle's location has a certain ambiguity, and the momentum is uncertain, but that is also limited. There is a difference between the crisp step function of the isolated pulse of Figure 2.04.2 (above), and the gradual transition edge of the pulse of the figure (below). This difference indicates that the voltage of the waveform is the probabilistic indicator of the elementary particle's position, as I will discuss later in sections 2.09 and 2.10.

The graphic fermion and boson identification has three consequences. The first is that there can be only two distinct types of elementary particles, fermion and boson. There is no other type of elementary particle. To build any structure, I need the materials that occupy their own spaces, and the media of assembling the materials—a fermion, which occupies its own space and refuses invasion of any other fermion—and the media of assembling the structure is executed by a boson. To make a structure, a fixed number of fermions are required, but various number of bosons are required.

The second is that the waveforms of fermion and boson can never be symmetrical. There is no way to make a natural-looking boson companion of a fermion from the boson waveform, and vice versa. So there can be no supersymmetry (SUSY) between them. This feature agrees with the futility of efforts to search for supersymmetric particles since 1990s. From the boson waveform, it appears that it takes the time of the pulse width of a boson to interact with fermion. Actually a boson waveform is an AND composite of a fermion and an antifermion waveforms, as shown in Figure 2.04.3(a). If a fermion-boson interaction is considered as structure building, the existing structure must first be removed by the antifermion, and then a new structure is built by the fermion. Then, this feature explains the time required for fermion-boson interaction, although the time can be infinitesimally short.

The third consequence is that the fermion's waveform is connected to the source of the step function by a structure. Because of this structure, there can be only one fermion in a single state. The fermion is not free; it is bound to its source. This is a new insight that emerges naturally from the digital equivalent circuit model. Isn't it rational to consider, if two objects cannot occupy the same space, that the first object has its own structure already occupying the space? In textbooks, this feature of the fermion is explained mathematically by the antisymmetrized wave function (W. Heitler, *Elementary Wave Mechanics*, Oxford, 1945). In the equivalent circuit model, this feature naturally emerges from the fermion's waveform. Each fermion has its own connection to its source, which I call the umbilical cord. A fermion is identified by its source. The umbilical cord is not observable by any external observer but is recognized by the other fermions. This prevents two fermions from going into a single state. Actually, a pair of fermions having opposite spin can go into a single state. This can be explained by working out more details of the equivalent circuit model including spin (section 3.08).

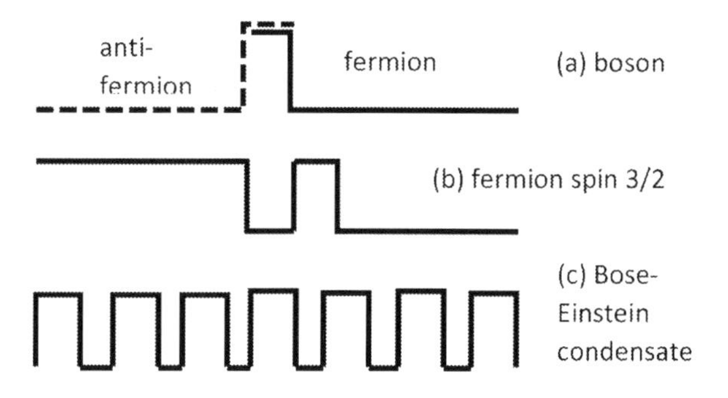

Figure 2.04.3 Composite particles

Figure 2.04.1 shows the simplest elementary particle waveforms in the equivalent circuit model. General waveforms can be constructed as a composite of the simple waveforms of a fermion and a boson in various ways. First, a boson is an AND composite of a fermion and an antifermion as shown in Figure 2.04.3(a). This is the case of a boson like meson, which is the composite of a quark and an antiquark. This feature explains the mechanism of the strong nuclear force. The second example of Figure 2.04.3(b) is the composite fermion having spin angular momentum higher than $\hbar/2$. The details of the fermion spin are discussed in sections 3.07 and 3.08. The last example of Figure 2.04.3(c) is stacked-up bosons in a single state. This is the state of Bose-Einstein condensation, as observed in the low temperature liquid helium or laser beam. From these examples, the step-function and isolated pulse are cleanly corresponding to the objects in the quantum world.

2.05 Fermions, Bosons, and Spin

In the universe's history, how did fermions and bosons emerge? In the phase 2 universe (section 1.06), buffer, logic gates, and capacitance emerged. The capacitor was the media of retaining the particle's energy. The particle and its propagating path were created. The model particle began to propagate the buffer chain. The particle's waveform was generated by the launcher (section 3.01), which was a latch that was a closed loop of NOR gates. When the latch was set, it sent out a step function, the model of fermion and antifermion. A pair of fermion and antifermion waveforms went through an AND gate, and the isolated pulse, the boson waveform, was created. Fermions emerged first, and then bosons were created from them. The pair of the particles was enough to describe any structure in the quantum world.

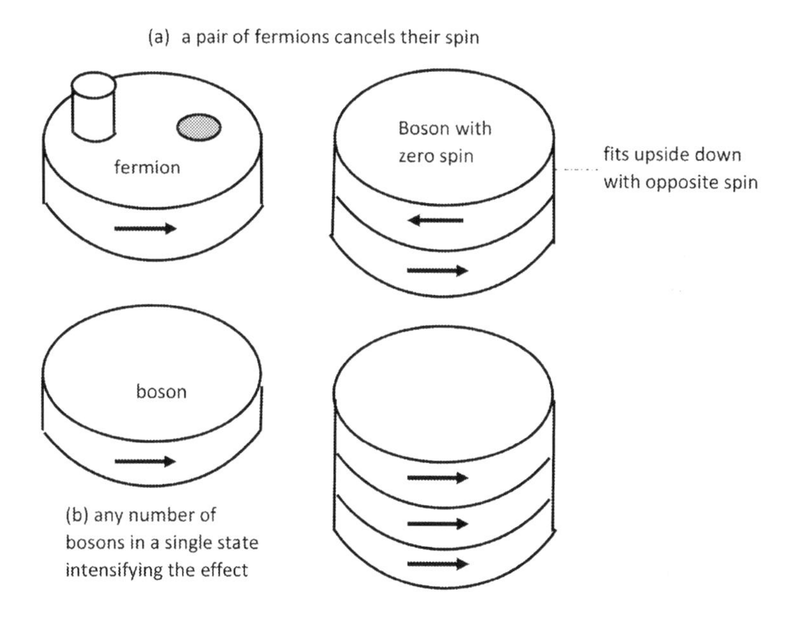

Figure 2.05.1 Composites of fermions and bosons

Fermions and bosons were distinguished by their statistical behavior; only one fermion pair having opposite spin could occupy a single fermion state. That can be modeled by a disc having a mutually connecting feature or not, as shown in Figure 2.05.1. Fermions and bosons can be distinguished by their angular momentum, the spin. Elementary fermions had spin $\pm\hbar/2$, and elementary bosons had spin $\pm\hbar$ or 0. The particle rotated and created the angular momentum (section 3.08). Where did this rotation come from? The launcher of the particles was a latch, a closed loop of NOR gates (section 3.01). The digital signal first rotated around the NOR gate loop to set the latch. This rotation was taken out from the latch and created the particle's angular momentum and its umbilical cords (section 3.08). For a fermion having only one loop, and the loop carried spin $\pm\hbar/2$, and a boson had a pair of loops, that carried spin $\pm\hbar$ or 0. The equivalent circuit model shows that the spin is independent from the particle's mass. It belongs to the loop structure that the particle makes (section 3.08).

How does this rotation affect the statistical behavior of the particle? A single state of a fermion is actually occupied by a pair of fermions having opposite spin. The model of the fermion is shown above Figure 2.05.1(a). When a pair of fermions having opposite spin joins and cancels the rotation, the spin angular momentum, the pair appears not to be rotating. They make a flat disc like a coin. Since fermions are the building blocks of the structure, it must be basically static. The pair's model is a rotating disc having the structure that fits together, that is, the connection of the spin-up and spin-down pair. The connection makes a static disc. As for the boson, any number of bosons can occupy a single state, like stacking up coins, and their effects are added up as shown in Figure 2.05.1(b). A boson must have the intensity parameter to execute the fermion-boson interaction. The boson's intensity is the number of stacked-up bosons Quantum world is built up by fermions and bosons.

2.06 Equivalent Circuit Model of a Fermion

A fermion has a connection between the particle and its source, the fermion launcher. The role of this connection is to assure that the present state of the fermion is not invaded by any other fermion. As the fermion travels through the quantum space, its trace is left behind and recognized by the other fermion so that it does not get into the same state. The connection between the fermion launcher and the fermion is created by the activity of the traveling fermion's substance. According to elementary particle physics, a fermion is surrounded by virtual bosons, that split into virtual fermions and antifermions. They screen the fermion's charge. As the fermion moves, the structure moves, while leaving a connection between the source and the fermion. The structure is not observable by the outsider. They are a pair of go-and-return buffer chains. The paired paths are required because when the present fermion's state is terminated, the connection and the launcher must be wiped out by sending the state termination signal to the launcher. Neither the launcher nor the buffer chain node is capacitor loaded. The structure is not observable, yet it is recognized by other fermions.

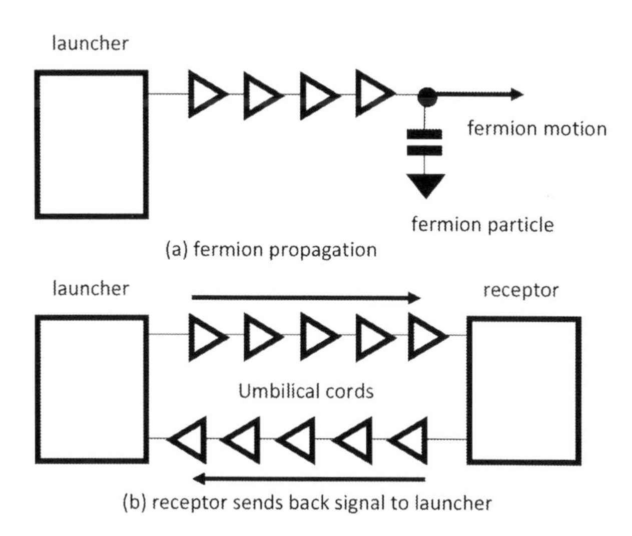

Figure 2.06.1 Structure of a fermion umbilical cord

At the end of the buffer chain (the present fermion location), there is a charged capacitor, whose energy is the substance of the fermion, as shown in Figure 2.06.1(a). A capacitor exists only at the end of the connection, that is, at the present location of the fermion, to hold its substance (mass or energy). This buffer node carries the fermion. As the fermion moves, the capacitor moves, and there is no capacitor-loaded node behind the present fermion location. The pair of connections are the fermion's *umbilical cord.*

The present fermion state is terminated by emergence of a fermion receptor at the time of its observation. When the receptor emerges, the launcher and the umbilical cord must be wiped out, so that another fermion is able to get into the state. For that purpose, the state termination signal must be sent to the launcher from the emerged receptor, as shown in Figure 2.06.1(b). The signal from it erases the launcher and the existing umbilical cord.

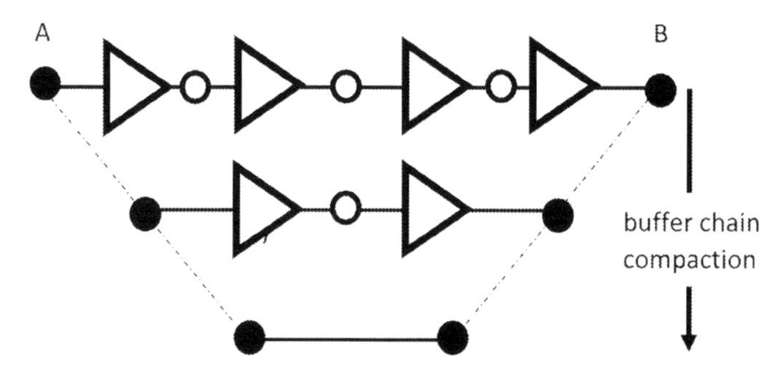

Figure 2.06.2 Compaction of gates

As the fermion's state is terminated, the fermion goes into a new state. A new fermion is launched by the emerged receptor. The launcher and the receptor must have the same logic circuit (section 3.01). At the moment of the present state termination, the pair of connections between the launcher and the receptor are both cascaded buffer chains whose nodes are not capacitor-loaded. The launcher and the receptor are symmetrically connected as shown in Figure 2.06.1(b) by the pair of the umbilical cords. Since no node of the umbilical cord and the launcher is capacitor-loaded, the signal terminating the fermion's state propagates at infinite velocity. This may appear to violate the relativistic requirement of locality, that no *information* can propagate faster than the velocity of light. Many authors wonder why such a velocity is achieved. Let us consider the state of the launcher. As long as the present state is maintained, the launcher's latch is set. This state is not an observable state since none of the node is capacitor-loaded; the launcher is set at the logic

HIGH level with probability 1. So are the states of the umbilical cords. Then the signal exchanged between the receptor and the launcher is the probability setting signal that may have infinite velocity (section 1.04). Figure 2.06.2 explains how such a velocity is achieved in the equivalent circuit model.

Stages of the cascaded buffer chain can be *compacted* to a small number of cascaded buffers and then to a direct connection, if none of the buffer's node is capacitively loaded. The distance between location A and B can be quite long in the quantum entanglement experiment, as the effect was observed to exist by the experiment proposed by John Bell (K. W. Ford, *The Quantum World*, Harvard University Press, 2004). A question I expect is, if a buffer occupies length L, why does delay time L/c never become a part of the buffer delay time, where c is the light velocity? This is because the buffer without inductance is the quantum effect model that emerged in the phase 2 universe (section 1.06), before the emergence of inductance. The triode-capacitor model emerged *before* light velocity set the velocity of information transmission. The present quantum world still maintains this feature of the universe, if the action is limited to the probabilistic state. The quantum state is, as it is, not real but a probabilistic existence. The probabilistic state becomes real only if a particle rides on the state. The signal that affects the state is the probability-setting signal, that propagates at infinite velocity.

Yet this nonlocality feature bothered many authors. Some authors wrote that the quantum system makes a *wormhole* between the initial and final locations A and B of Figure 2.06.2. Fascinating, yet no evidence has been available to prove existence of the extra dimension in the present space. I prefer a mundane mechanism, that by some yet unspecified ways, the signal can go through the structure of the screening cloud made by the particles, by a mechanism similar to idealized falling dominoes (*Self-Consciousness*, iUniverse, 2020). A buffer chain unloaded by capacitance is able to model such structure. This means that the length of the buffer L in my model is a parameter that requires explanation later (section 3.06). The uncertainty principle $\delta x \delta p_x = \hbar/2$ is basic to this explanation.

A boson consists of a paired fermion and antifermion. Since the two particles jointly create a pair of connections in the quantum vacuum, the structures behind the particle cancel each other, and there is no connection of the boson to its source. What I call a boson's umbilical cord later in section 3.09 is the path of a pair of bosons between a pair of interacting fermions.

2.07 Fermion Replacement Motion

How does a fermion move in the space and create the umbilical cord? Figures 2.07.1 (a)–(e) show the mechanism of the cord creation. There is a fermion (black circle) at location 0 of Figure 2.07.1(a). This fermion makes a virtual fermion-antifermion pair at location 1, in the direction of its motion. In Figure 2.07.1(b), the gray circle indicates the virtual fermion and the white circle the virtual antifermion. The virtual antifermion moves back to location 0 in Figure 2.07.1(c), by the force working between it and the fermion at location 0. A particle and its antiparticle carry opposite charges and the force is attractive. The virtual antifermion carries no substance (energy or mass) and therefore the motion takes place instantly. At location 0, the fermion that was there (the black circle) gives away its energy and becomes a virtual fermion. This virtual fermion and the moved back virtual antifermion go into a pair annihilation in Figure 2.07.1(d). The energy released by the fermion travels from location 0 to location 1 and converts the virtual fermion there into a real fermion in Figure 2.07.1(e). This process is repeated to move the fermion in the right direction.

Steps of the process of Figures 2.07.1(a)–(e) create the pair of umbilical cords of the fermion and of the associated virtual antifermion in the equivalent circuit model, as shown in Figures 2.07.1(ab) and (cde). Here, (ab) and (cde) refer to the steps of Figures 2.07.1(a)(b) and (c)(d)(e), respectively. Creation of the pair of inverters by steps (a)(b) is not observable, and therefore both inverters are not capacitively loaded. The capacitor at node 0 still holds the fermion's substance, and the fermion is still there. These processes are followed by processes (c)(d)(e), where the process (d) generates energy at location 0. The energy travels from location 0 to location 1 and converts the virtual fermion to a real fermion. The processes create another pair of inverters to the upper and the lower umbilical cords. Since the energy must travel from location 0 to location 1, it requires time. Because this is the model of the phase 2 universe, the time is set by the load capacitor charging by the triode, and not by the delay time of light velocity through the buffer's length. Therefore only the second inverter of the upper umbilical cord is capacitor-loaded. The fermion at location 1 is now observable. The capacitor that existed before at location 0 jumped to location 1. The fermion moved from location 0 to 1.

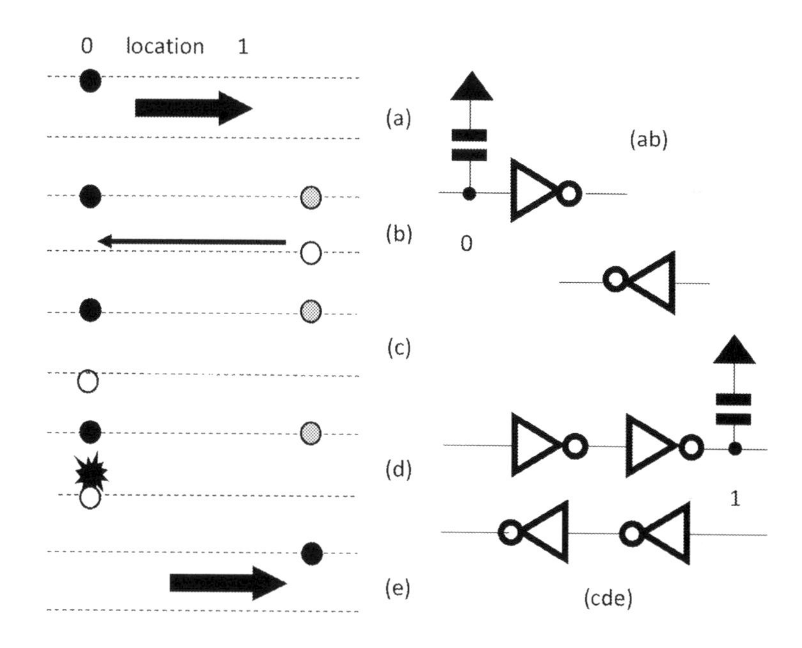

Figure 2.07.1 Equivalent circuit generation

After the processes (a)–(e), the capacitor at node 0 moves to node 1. After the emergence of capacitor at location 1, the charge that was in the capacitor at location 0 is used to charge up the capacitor at location 1. The location 0 capacitor decreases while giving the charge away to the location 1 capacitor. The capacitor *jumps* from location 0 to 1. This process is discussed in sections 2.09 and 2.10. The processes are repeated to move the fermion, and to build the pair of the fermions' umbilical cords. The upper umbilical cord models the buffer chain of the real fermion's motion. The lower umbilical cord models the associated virtual antifermion's motion, created and then annihilated. Then the upper buffer chain's logic level is HIGH (the fermion exists). In the lower buffer chain, the virtual antifermion does not exist; the particle is generated but is then annihilated. The antifermion's empty state is also logic HIGH level. When the fermion's state is terminated, the logic levels of the upper umbilical cord turn instantly to the logic LOW level at the signal from the fermion receptor resetting the latch of the launcher.

In this capacitor jump mechanism, the location 0 capacitor decreases while giving its charge to the location 1 capacitor. This means that location 0 has no more capacitor loading after the particle has moved from location 0 to location 1. The same thing happened in the nodes behind location 0 back to the fermion's source. Behind the fermion, no location has any capacitor loading. Such an umbilical cord is a probabilistic existence and is not observable, as I discussed in section 2.03, but is still recognized by the other fermions. The signal goes through such a buffer chain at infinitely high speed. I note here again that the buffer delay time does not include the light velocity delay time, since this is the phase 2 universe model. This is the basic mechanism of the umbilical cord: to maintain the fermion in its state and then to erase the umbilical cord instantly if the fermion's state is terminated. The model has only one charged capacitor. The fermion replacement mechanism that moves it may be considered as the conversion of a fermion to a boson (the virtual fermion-antifermion pair ahead of the fermion path) and back to a fermion later. A particle's internal symmetry (fermion versus boson) conversion creates the fermion's spatial motion. This is an interesting feature of this fermion propagation mechanism.

The same scheme works to build an independent umbilical cord of the particle's spin. A fermion's spin is the angular momentum of the particle, quantized to $\pm\hbar/2$. The direction of the spin, up or down, is specified by a single binary variable, HIGH or LOW. While the substance's umbilical cords are being built, the fermion and antifermion make their own spin umbilical cords, a pair of go-and-return buffer chains, which are independent from the substance's umbilical cord. The virtual fermion carries the same spin as the real fermion, and the virtual antifermion carries the opposite spin, to maintain the spin angular momentum conservation. Since no energy transfer occurs, a spin umbilical cord is not capacitively loaded. The spin-setting signal propagates the cord at the infinitely high velocity of the probability signal.

2.08 Boson Replacement Motion

A boson is a composite of fermion A and antifermion B, as shown in location 0 of Figure 2.08.1(a). A and B are not the same kind of fermion, since if they were, they would pair-annihilate right away. The boson moves as the pair of the fermion and the antifermion, and therefore the moving mechanism is a composite of the fermion pair's motion. This makes the model simpler than that of the fermion, by canceling the lower umbilical cord.

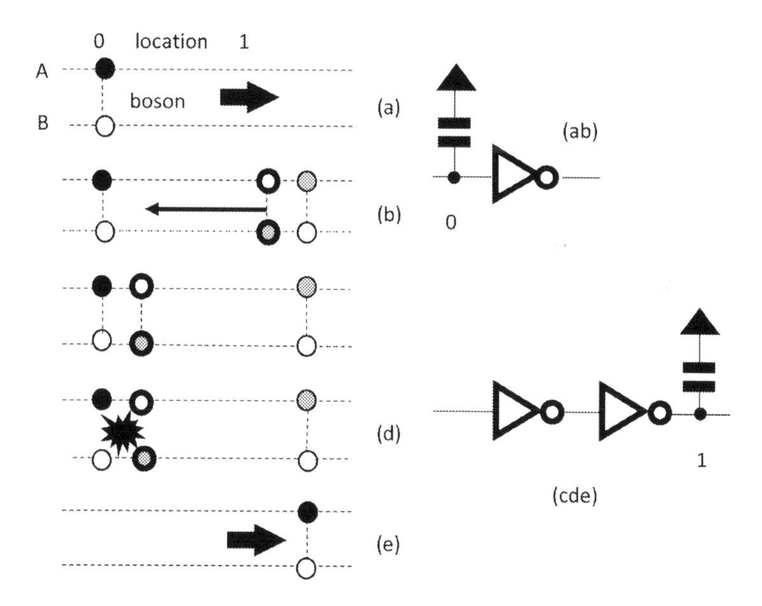

Figure 2.08.1 Equivalent circuit generation of boson propagation

Similar to fermion motion, the boson AB at location 0 makes the virtual boson (the virtual fermion pair) ahead of the path at location 1, and a virtual antiboson near location 1, as shown in Figure 2.08.1(b). Virtual antiboson is shown by the circle having a thick boundary. This antiboson at location 1 moves back to location 0 and converts the boson that was there to a virtual boson. Then the *pair annihilation* of the virtual boson and virtual antiboson releases energy, as in the fermion's case. This is a pair of virtual fermion-antifermion *annihilations* at location 0, generating energy. The energy travels from location 0 to location 1 and converts the virtual boson to a boson. The energy is sent from location 0 to 1 to convert both the virtual fermion and the virtual antifermion pair of the virtual boson at location 1 into a boson. By repeating this process, the boson moves to the right.

The process of Figure 2.08.1(a)(b) creates a stage of inverter as shown in Figure 2.08.1(ab). This virtual boson-antiboson creation is not observable, and the inverters are not capacitively loaded. By the process of Figure 2.08.1(d), energy is released at location 0 and moves to location 1. The operation is symmetrical with respect to the pair of fermion and antifermion making up the boson. Energy movement requires time, and therefore the second stage of the inverter of Figure 2.08.1(c)(d)€ is capacitively loaded.

The motion of the boson is a joint action of the fermion and the antifermion pair making up the boson. Since I consider the pair as a single particle, a single buffer chain, instead of the pair of buffer chains of the fermion motion of the last section, is adequate. Since boson motion involves the motion of the particle and antiparticle together, the boson at location 0 becomes the source of the boson at location 1 in every step of motion. Because of that, the boson is free from its real source. The process creates a single buffer chain and the buffer chain left behind the boson to its real source plays no role. The buffer output nodes behind the boson are all at logic LOW level, once the boson passes each location (section 3.03). Boson propagation leaves no trace behind the moving boson, and there is no umbilical cord connecting the boson to its real source. This is required, since the boson is a free particle that transmits force between any fermion pair. Yet when a pair of fermions interacts, that makes a pair of continuously moving go-and-return boson paths between the pair of fermions (section 3.09). That is similar to the pair of buffer chains connecting a fermion to the source. The interacting fermions keep exchanging bosons. That is similar to the umbilical cord of a fermion, and therefore I call that the boson's umbilical cord.

As for the spin of the boson, the fermion and antifermion making up the boson create a separate spin umbilical cord. This is because some bosons have spin 0 and ±1 Spin umbilical cords are required to assure the particle's spin angular momentum conservation. The mechanism of making each of the boson spin umbilical cord is the same as that of the fermion. Spin umbilical cord has significant role to explain the quantum entanglement (section 3.11).

2.09 Capacitor Jump Mechanism

The quantum vacuum should not spend free energy to move a particle from one location to the other. This feature should be reflected in the equivalent circuit model, because the equivalent circuit is not a classical mechanical model that conserves free energy when it operates. The circuit generates heat and is affected by it. Then how does a fermion propagate in the equivalent circuit model? Figures 2.09.1(a)–(c) show the mechanism. A buffer chain section made of the three inverters INV1, INV2, and INV3 is shown in Figure 2.09.1(b). INV2 and INV3 make a single buffer stage. The accessible nodes N1 and N3 are shown by closed circles. Figure 2.09.1(a) shows the inverters INV1's and INV3's circuit diagram, showing the crucial operation of the PFET MP1 and MP3 of INV1 and INV3, respectively. Nodes N1, N2, and N3 waveforms are shown in Figure 2.09.1(c). Suppose that node N1 capacitor C_0 of Figure 2.09.1(a) is fully charged by the current of PFET MP1 of INV1 and the voltage reaches V_{DD} as shown in Figure 2.09.1(c). At that moment, inverter INV2 makes an instantaneous HIGH to LOW logic level transition, since node N2 is not capacitively loaded. The PFET MP3 of INV3 begins to charge the node N3 capacitor C_0 of Figure 2.09.1(b), and it reaches the fully charged voltage V_{DD} as shown in Figure 2.09.1(c). Node N1 and N3 waveforms are exactly the same, if the time is shifted.

This process is repeated to propagate the fermion's step function waveform in the right direction. Here, I consider that capacitor C_0 of node N1 has returned its entire charge to the power supply and made a jump to node N3. This capacitor was and is the only charged capacitor in the entire buffer chain to hold the energy of the fermion. The energy must be temporarily stored. For that purpose, the power supply, the quantum vacuum, must have a capacitor to keep the charge and energy of capacitor at node N1. But the mechanism does not appear natural. This is the problem of this model, called the *semiclassical* equivalent circuit model (in spite of this problem, this model has easy graphic presentation, and I use that later). How can I modify this feature to model the quantum phenomena more naturally? I reinterpret the model operation to make it natural as a quantum mechanical process. I try to come up with a real quantum mechanical model, by reinterpreting the mechanism involved in the semiclassical model.

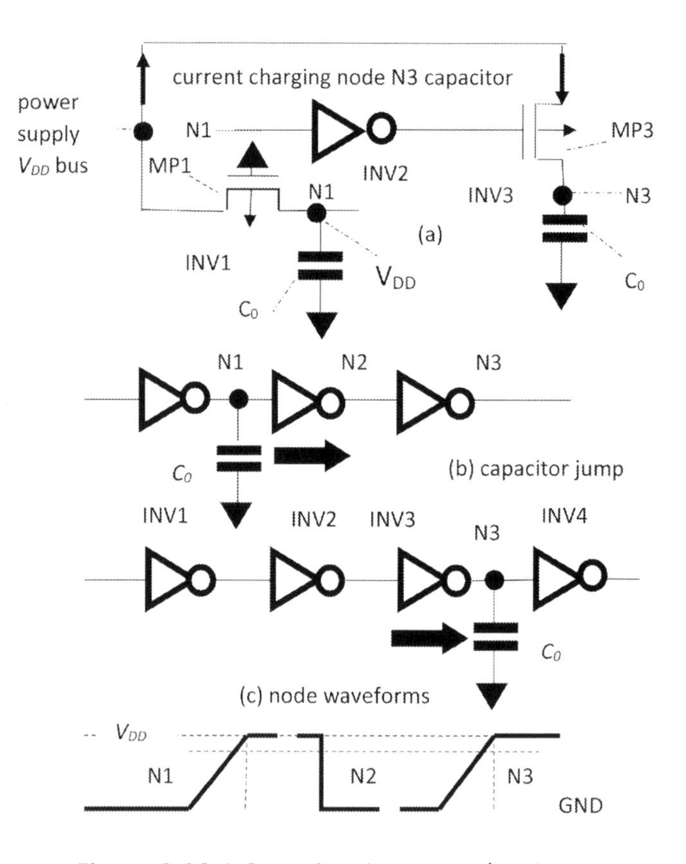

Figure 2.09.1 Capacitor jump mechanism

When node N1 reaches voltage V_{DD}, the capacitor C_0 there is fully charged. Its voltage equals the power supply voltage V_{DD}. The fermion is there. At that moment, a small capacitor emerges at node N3 and it begins to charge. Where does the charge come from? The quantum vacuum, in which the circuit is placed, should not give or take any charge, just for the fermion to move. I mentioned before that node N1 capacitor C_0 returns its entire charge to the quantum vacuum. To execute this process more realistically, capacitor at node N1 *decreases* while returning the charge to the power supply, the quantum vacuum. Capacitor C_0 at node N1 becomes a part of the power supply, while the capacitance decreases from C_0 to zero. Along with node N1 capacitor's decrease, node N3 capacitor increases from 0 to C_0. How does this happen? The node N1 capacitor voltage remains infinitesimally above V_{DD} as the capacitor decreases. The PFET MP1's differential conductance in this state is infinity, as I assumed in section 2.03. At the same time, the node N3 voltage is infinitesimally lower than V_{DD} as the node N1 capacitor charge flows into the small node N3 capacitor. The two mechanisms work together.

The two effects, first an infinitesimal decrease of the local power supply voltage from V_{DD} by charging node N3 capacitor and second the infinitesimal increase of the local power supply voltage from V_{DD} by decreasing of capacitor at node N1, together cause a continuous flow of charge first from the capacitor at node N1 to the power supply bus and then to node N3 capacitor as shown in Figure 2.09.1(a). In this process, node N3 voltage is only infinitesimally lower than V_{DD}.

MOSFETs are source-drain symmetric devices, and current can flow both from the power supply to the node capacitor or from the node capacitor to the power supply. This is a reversible (quasi-static) process that does not increase the model's entropy. This is why I say that the charged capacitor at node N1 becomes a part of the power supply. In the end, the node N1 capacitor vanishes. Not only the particle energy, but also the node capacitor moves continuously from node N1 to node N3 for the particle to move from node N1 to node N3. This is the crucial point of the reinterpreted quantum mechanical model: the particle's energy *follows* the capacitor's motion. I mentioned before that the quantum state is not made each time by the articles, but it does exist as the asset of the quantum vacuum. Such a preexisting state works as I just described.

Similarly, all the buffer nodes behind node N1 to the fermion's source gave away the charges, energies, and the capacitors. Yet all the nodes between the fermion and its source are kept at the logic HIGH level V_{DD}, because the fermion's source (launcher) maintains the logic HIGH level (section 3.01). The launcher's support maintains the fermion's umbilical cord at the logic HIGH level. Although these nodes have no capacitor and therefore are not observable, they prevent the other fermion from invading the state. The buffer chain, whose nodes are not capacitively loaded, is unobservable, but it transmits the probability setting signal at infinitely high velocity. This mechanism works to wipe out the umbilical cord and the fermion launcher, when the fermion's present state is terminated by observation.

Another detail of the model is, when the charged capacitor at node N1 decreases and returns the charge and energy to the power supply by reducing the capacitance, a force is exerted to separate the pair of plates of node N1 capacitor to reduce its capacitance from C_0 to 0. Where does the force come from? During that time, the node N3 capacitor that increases from 0 to C_0 is charged. The capacitor was originally small, having a wide electrode gap. The node N3 capacitor plate electrodes are pulled together by the electrostatic force of the charge from the node N1 capacitor. To increase the capacitance of the node N3 capacitor from the originally small value to full value C_0, the pulling force must be balanced somewhere. The force that pulls the node N1 capacitor plates apart to reduce its capacitance is balanced by the force that pulls the node N3 capacitor plates together.

Let us observe the switching process of the circuit model. The node N3 capacitor was originally small, but a small amount of charge is transferred from node N1 to node N3. The nod N3 voltage, even at the very beginning of the capacitor transfer, is only infinitesimally lower than V_{DD} to charge the originally small node N3 capacitor. Since the current flows from the node N1 capacitor, whose voltage is infinitesimally higher than V_{DD}, to node N3 capacitor, whose voltage is infinitesimally less than V_{DD}, this is thermodynamically a quasi-static process, that keeps the circuit's entropy basically unchanged. The node N3 capacitor depends on time, and I write it as $C_0(t)$. Quantum mechanically, the node N3 voltage must be referred to the full capacitance C_0, since the node capacitor's energy must represent a *not dividable* elementary particle's substance, its energy. Then, the *effective node N3 voltage* in the semiclassical model depends on time, such that $V_{N3}(eff) = V_{DD}C_0(t)/C_0$, and when the node N3 capacitance is fully charged, $C_0(t) \rightarrow C_0$, and the node N3 voltage $V_{N3}(t) \rightarrow V_{DD}$, that is, $V_{N3}(eff) \rightarrow V_{DD}$.

The effective node N3 voltage increases linearly with time (section 2.03), but this voltage does not have proper physical interpretation. Circuit-theoretically, the linearly increasing node voltage $V_{N3}(t)$ is real, but quantum-mechanically it is not. Why not? Because the elementary particle is not dividable any further, and the capacitor C_0 and its node voltage V_{DD} are both the determining factor of the not dividable particle's energy. The circuit-theoretical node voltage becomes the *probability of existence* of the particle at node N3. This interpretation makes the semiclassical model and quantum-mechanical model compatible.

2.10 Details of the Capacitor Jump Mechanism

The *capacitor jump* is the key mechanism of the quantum effect modeling by the CMOS digital equivalent circuit. By the capacitor jump mechanism, quantum mechanical features are brought into the simple semiclassical digital equivalent circuit model. So I reinterpret the mechanism by looking at every step of the model circuit's operation once again, and explain its physical significance. I refer to Figure 2.09.1 of the last section as the reference.

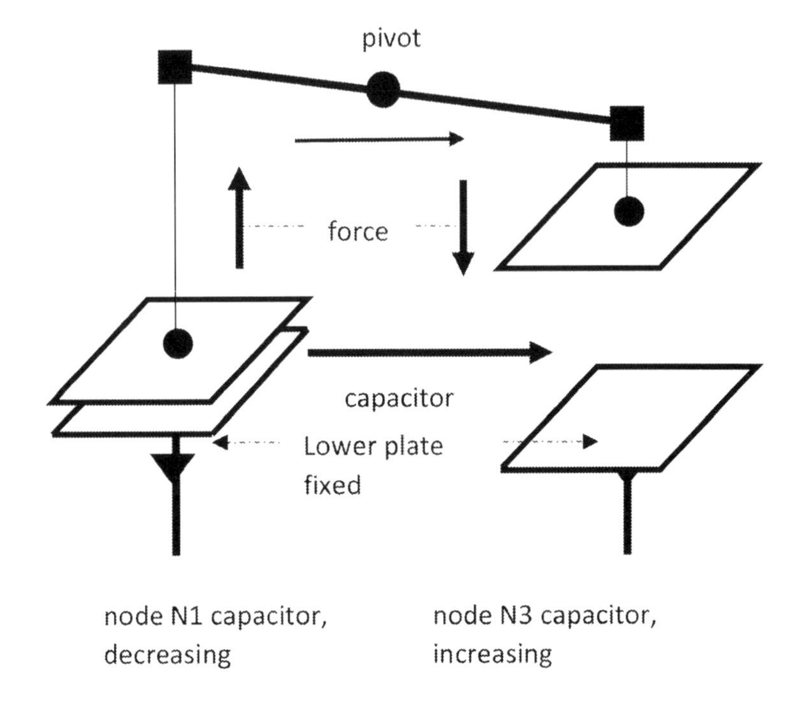

Figure 2.10.1 The force that changes the node capacitance

By the capacitor jump mechanism, the capacitor of node N1 decreases while the capacitor of node N3 increases. Finally the node N1 capacitor becomes zero, and node N3 capacitor becomes C_0. The node N1 capacitor's charge and its energy move together with the capacitor to node N3. That models fermion motion from node N1 to N3. This model's operation is quite different from the conventional equivalent circuit model operation, in which the model circuit component or connectivity never changes during its operation. A feature of a self-altering circuit is used in this model. The digital circuit is also self-altering among the gate connections, but in the capacitor jump mechanism the load capacitor movement is the key feature of the circuit operation.

The human brain is one such self-altering circuit (*Self-Consciousness, Human Brain as Data Processor*, iUniverse, 2020). The capacitor jump mechanism originated in my mind from the similarity of the brain function to a quantum system's operation. In the human brain, the connectivity of neurons increases in the high-activity region of the brain. The high-activity region tends to gain more processing elements and connections than the low-activity area, and this is the similarity between the human brain and the quantum capacitor jump mechanism. I review this mechanism once again.

The node N3 gets small capacitance and small charge from the node N1 capacitor when the capacitor jump begins. The node N3 voltage is then already very close to V_{DD}, and is only infinitesimally lower than that, so that the current can flow from node N1 to node N3 via the power bus (Figure 2.09.1). The node N1 voltage is only infinitesimally higher than V_{DD} since the charge is lost, but the capacitance decreases even more. The current first flows from the node N1 capacitor through the PFET MP1 channel of the inverter INV1 to the power supply bus, and then through the PFET MP3 channel of INV3 to the node N3 capacitor. Both PFET channels have infinite differential conductance up to current I_0 (section 2.03). Then the charge transfer is a quasi-static thermodynamic process.

As more capacitance and more charge are transferred from node N1 to node N3, the energy of the capacitor at node N3 reaches finally the full fermion energy $C_0 V_{DD}^2/2$, and that much energy is lost from node N1. During that time, the node N1 capacitor decreases from C_0 to 0. As all the energy and capacitor of node N1 are transferred to node N3, the fermion moves from node N1 to node N3. In this state, node N1 has no more capacitor. The only capacitor is now at node N3, holding the fermion's substance, the energy. The same process occurred at all the nodes behind node N1.

The charge is transferred from node N1 to node N3, but how is the capacitor transferred? Figure 2.10.1 shows the mechanical model of the capacitor transfer. A capacitor consists of a pair of plate electrodes. The plate size is the same at all the nodes. Then the closer the gap is between the plates, the higher is the capacitance. The capacitor transfer is effected by the electric force working between the pair of capacitor plates. The lower capacitor plates are fixed to the GND, and the upper plates move during the capacitor transfer.

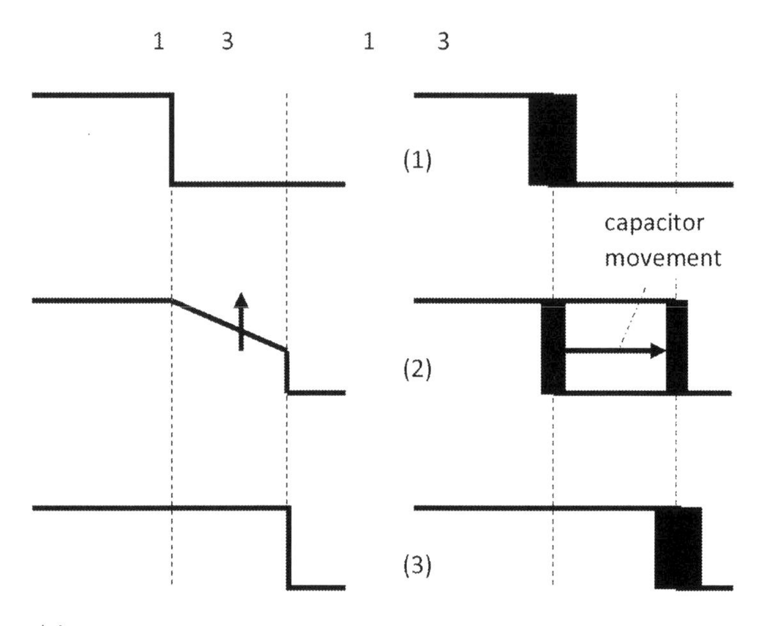

Figure 2.10.2 Semiclassical and quantum capacitor jump mechanism compared

The upper capacitor plates are connected to the mechanical balance of Figure 2.10.1. The node N3 capacitor plates pull closer and closer together by electrostatic force as it is more and more charged. This force pulls the node N1 capacitor plates apart, to reduce the node N1 capacitor. To match the forces, the pivot location of the balance must move from left to right in Figure 2.10.1. In this way the node N1 capacitor decreases and the node N3 capacitor increases. In the end, there is no capacitor at node N1. The only capacitor C_0 is at node N3, which now carries the energy of the only fermion in the propagation path. All the nodes behind the fermion's location lose capacitance by the same mechanism, and it is converted to the buffer chain that transmits a probability setting signal at infinite velocity. This buffer chain carries out the only role to erase the trace of existence of the fermion in the previous state, by propagating the probability setting signal to erase the umbilical cords. This is the new feature of the quantum mechanical model.

Let us compare this mechanism with the simple *semiclassical* capacitor jump mechanism that I have been considering since section 2.03. This new model was derived from the semiclassical propagation model I described before (*Self-Consciousness, Human Brain as Data Processor*, iUniverse, 2020). In that model, the capacitor at node N1 gives all the charge away to the power supply at the very beginning of the capacitor jump. It becomes uncharged and then jumps instantly from node N1 to node N3 (that is possible because the capacitor carries no energy). Then the node N3 capacitor begins to charge by the current from the power supply bus. This is a simple mechanism to understand, yet it has several unnatural features. This model assumes that the power supply bus must have a capacitor to hold the charge from the capacitor of node N1 temporarily, before delivering it to the node N3 capacitor. This is an extra and artificial model feature. Node N3 voltage goes up linearly with time, from 0 to V_{DD} during the switching time, $C_0 V_{DD}/I_0$. The node N1 capacitor had already given the charge away and vanished in the semiclassical model. How does this capacitor movement from node N1 to N3 happen? That was the point that has never been explained by the semiclassical model. I needed to explain how the capacitor jumps (it *can* jump, but *how* has not been explained) and what the continuous node N3 voltage of the simple semiclassical model physically means. These are the key part that this new quantum mechanical model is able to explain.

In the reinterpreted quantum mechanical model (I call this new interpretation by this name), the node N3 voltage of the model circuit is always infinitesimally close to V_{DD} during the energy-capacitance transfer, but the node capacitance increases from 0 to C_0 during the gate's switching time. The two models' node voltage developments have different physical interpretations, as schematically shown in Figure 2.10.2. Figure 2.10.2(a) shows the semiclassical model. In Figure 2.10.2(b) (1), (2) and (3), the thickness of the voltage transition line shows the capacitance values of node N1 and N3. The node N3 voltage is already very close to V_{DD} from the very beginning of the capacitor jump. If I compare the energy of node N3, the two models give the same result. The two models are physically equivalent, but the physical meaning of the circuit model's node voltage, that is, the effective node voltage $V_{N3}(eff)$ of the last section, is not a real voltage, but it gives the measure of probability of the fermion's existence in node N3 and nonexistence in node N1 in the quantum mechanical model.

In spite of the model's apparent similarity, the physical meaning of the node voltage is quite different. The gradually increasing effective node voltage of the simple semiclassical capacitor jump model carries clear circuit-theoretical meaning, but does not have real physical meaning in the quantum mechanical model. In the quantum mechanical model, the voltage becomes the indicator of the probability of finding the particle at either node N1 or N3. The particle is surely in node N3 only if the node N3 voltage is V_{DD} and the capacitor there is fully at C_0. This is because the elementary particle cannot be divided. Only the probability of existence of the particle is indicated by the simplified semiclassical model's gradually increasing effective node voltage. If the node N3 voltage is not V_{DD} or the node N3 capacitance is not C_0, the particle is still not yet in node N3. This observation leads to the clear definition of buffer length used in the model. The length must be chosen as the width of the wave packet derived from the linear circuit model. This is a key connection between the linear and the digital circuit models.

By this modification and reinterpretation, the probabilistic feature of the quantum phenomena is introduced into the simplistic digital semiclassical equivalent circuit model, by giving a different meaning to its node voltage. In addition, the quantum mechanical model explains the capacitor jump mechanism explicitly. Why the capacitors jump and why the capacitors vanish behind the particle's location are now clearly explained. The semiclassical digital circuit model must have this reinterpretation of the capacitor jump mechanism and also the reinterpretation of the node voltage appended to it to make it the real model of the quantum phenomena.

In spite of this defect of the semiclassical model, I still use the semiclassical model in the later part of this book. The reason is, it is practically impossible to draw a comprehensible illustration of the capacitor jump of the quantum mechanical model. Figures such as Figure 2.10.2(b) only confuse the reader. Yet if the node waveforms of the semiclassical model are properly interpreted as I discussed here, the semiclassical model is still easily comprehensible from a simple drawing.

Yet another unexplained detail of the model is as follows. While the capacitor at node N3 is charged, its voltage is not exactly V_{DD} but infinitesimally lower than that. The current flowing through the PFET channels of INV1 and INV3 of Figure 2.09.1 of the last section generates heat in the PFET channel. Although the heat is small because the process is thermodynamically quasi-static in the reinterpreted quantum model, it still consumes a small amount of the free energy. Does the quantum vacuum spend even that small free energy to move the particle? No. The reason is that the environment of the equivalent circuit, the quantum vacuum, sends the heat energy back to the circuit's power supply as free energy. For the equivalent circuit model to be consistent, its environment, the quantum vacuum's thermodynamic character, must allow that. I need to define the required thermodynamic character of the quantum vacuum. The energy apparently emerging from the quantum vacuum fluctuation is enormous. But the energy emerging from it is always taken back. The perpetuum mobile of the first kind is impossible to be built. Then, could the quantum vacuum be the high-temperature source of the perpetuum mobile of the second kind? For that purpose, the temperature of the associated heat sink must be lower than the temperature of the quantum vacuum. But where is the heat sink? There is nothing except the quantum vacuum itself. The perpetuum mobile of the second kind is also impossible. Then the quantum vacuum's temperature must be absolute zero.

I assume this thermodynamic character of the quantum vacuum as the required environment of the digital equivalent circuit model. By giving heat energy to the quantum vacuum, it sends back the same amount of free energy to the circuit. This is the same thing as that the temperature of the vacuum is absolute zero, by the second law of thermodynamics. Any heat generated by the model circuit drives a 100 percent efficient thermal engine, which drives the electrical power generator to supply the free energy back to the circuit. I *assume* this kind of quantum vacuum as the circuit's environment. This circuit environment makes the model entirely consistent.

When the fermion goes through a state change, the charged capacitor jumps from the input of the logic circuit of the receptor-launcher to the receptor-launcher's output, because there is no capacitor-loaded node between the input and output locations. The circuit between the two locations work instantly (section 3.01). The fermion state change is effected by a logic gate:

by an OR gate for the fermion absorbing the boson, and by an exclusive OR gate for the fermion emitting the boson (section 3.01). The logic gate and the fermion launcher that follows the logic gate should have no capacitive load. Such a logic gate structure (the model of the quantum state transition) existing in the quantum vacuum specifies that their nodes should not be loaded by capacitor, so that the capacitor jumps across the circuit. The capacitor jump mechanism across the receptor-launcher plays the role of maintaining conservation of the number fermion at the time of nuclear reaction in the equivalent circuit model.

2.11 Parameter Calibration including Inductance

In section 1.07 I mentioned that calibration of the equivalent circuit element's parameters took place when inductance emerged. That occurred at the shift from the phase 2 to phase 3 universe, when the present space-time measures were established by emergence of the basic physical constant *c*, light velocity. To set up light velocity as the basic parameter, inductance is required. I show a simple case, how the calibration affected the buffer drive current to meet the requirement of the new regime, compatibility with relativity. By inclusion of inductance, the time and spatial measures changed, and the phase 2 universe model parameters must be calibrated to the present phase 3 universe parameters.

Figure 2.11.1(a) shows the triode-capacitor model of the phase 2 universe. Let the load capacitor carrying the particle's energy be C_0, the buffer drive current be I_0, the power supply voltage be V_{DD}, and the buffer length be L. Then the velocity of the particle is given by LI_0/C_0V_{DD}. This is the velocity of the particle carrying energy $C_0V_{DD}^2/2$, and therefore the velocity must be less than the light velocity *c* in the phase 3 universe. This is the requirement from relativity that had not been satisfied in the phase1 and 2 universe, where the signal was all probabilistic. Now the following inequality holds.

Particle velocity = $LI_0/C_0V_{DD} < 1/\sqrt{\varepsilon\mu} = c$ (1)

where the right side of the inequality sign < is the light velocity, where ε and μ are the space's dielectric and magnetic susceptibilities. I assumed that the capacitor C_0 has square shape having edge length L, and the gap of the electrode is λ. Then, capacitance $C_0 = \varepsilon L^2/\lambda$. As in section 1.06, the triode is much smaller than the capacitor. Then the capacitor size sets the size of the buffer. Let us consider the first-order correction of the buffer drive current I_0 by including the inductance.

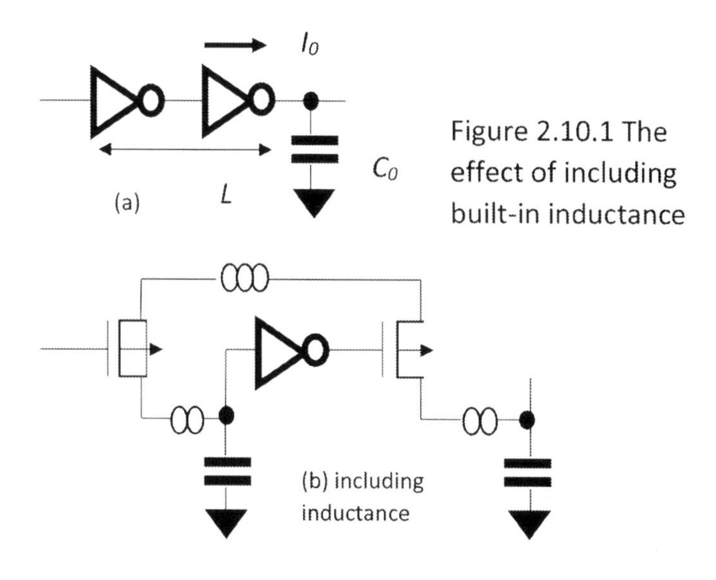

Figure 2.10.1 The effect of including built-in inductance

By simply rewriting Eq. (1) to set the velocity limit at the light velocity, I get

$$(L/\lambda)(V_{DD}/I_0) > (\mu/\epsilon)^{1/2} \quad (2)$$

where $(L/\lambda) > 1$ as I assumed in section 1.06 (the capacitor's plate size is larger than the plates' gap). In Eq. (2), the right side is the impedance of the free space. Since (V_{DD}/I_0) is the effective impedance of the buffer, the left side of eq. (2) must be much higher than the vacuum's impedance to satisfy the relativistic requirement. This requirement means to reduce I_0, and that is effected by inserting inductance to all the components and the connections of the buffer chain having nonzero size, as shown in Figure 2.10.1(b). All the inductances between the pair of the node capacitors are series-connected. By including the inductance, the voltage change between a pair of consecutive capacitor nodes does not create current between the nodes immediately after the first node's voltage change. It takes the current buildup time through the inductance. This is, in effect, to reduce the current I_0 between the nodes.

How much does the inductance effect modify the buffer drive current? The capacitance C_0 does not change, because I consider the first-order correction to the relativistic model. As I show in section 3.04, C_0's correction by velocity is of the order of $1/c^2$ which is small in this first approximation. Yet the buffer drive current changes by inclusion of inductance. Let the buffer drive current at phase 2 and phase 3 universe be identified by the extra suffixes 2 and 3. By inclusion of the effects of light velocity c, the light velocity signal delay is added to the triode-capacitance delay as follows:

$$(C_0V_{DD}/I_{02}) + (L/c) = (C_0V_{DD}/I_{03}) \text{ or } (1/I_{03}) = (1/I_{02}) + (L/cC_0V_{DD}) \quad (3)$$

where I_{03} in the model of phase 3 universe decreases from that of phase 2 universe. Then, the particle's velocity through the buffer chain ultimately cannot exceed the light velocity c. Since in this approximation C_0 wea assumed unchanged, the particle velocity decreases by inclusion of inductance by decreasing $I_{0.}$ The nonrelativistic quantum mechanical model in the present phase 3 universe uses this calibrated drive current, but does not include inductance explicitly in the model. By taking the inductance effects into account, the parameters of the phase 2 equivalent circuit of Figure 2.10.1 are calibrated to conform the relativistic requirements in the present phase 3 universe, while keeping the same equivalent circuit model consisting of triode and capacitor only, with no inductance. Equation (3) shows how inclusion of inductance calibrates the buffer parameters to give the lower particle velocity than light velocity. Yet the velocity of the probability signal still remains infinitely high.

3

QUANTUM PHENOMENA

3.01 Fermion and Boson Launcher

A fermion's umbilical cords consist of a pair of go-and-return CMOS buffer chains between the source and the fermion. The cords require a source that launches a step-function, the model of the fermion. The fermion launcher is a set-reset latch, that maintains the set state until the fermion's state is terminated. The latch is made from a closed loop of NOR gates (section 2.03), as shown in Figure 3.01.1. The charged capacitor shows that the fermion particle jumps across the logic gate and the launcher's digital circuit.

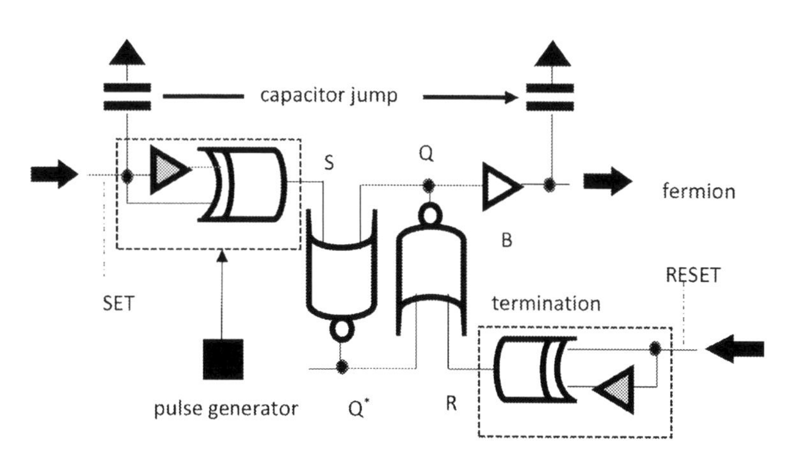

Figure 3.01.1 Fermion launcher

The latch is set or reset by a narrow upgoing pulse, that is generated by a combination of a buffer that creates the minimum delay time (shown by the gray color) and an exclusive OR gate (section 2.03). The circuit in the dotted box of Figure 3.01.1 is shown later by a black square box for simplicity. The circuit generates a minimum width upgoing pulse when the input signal's logic level changes. The pulse width is the minimum resolution time (section 1.06). Since no node of the set-reset latch is capacitively loaded, the latch is set or reset by the minimum width pulse. The first capacitor-loaded node in the downstream is the first section of the new fermion's umbilical cord B. The charged capacitor arriving at the launcher's input, which models the incident fermion, jumps to the buffer B's output. A new fermion is launched from node B to the new buffer chain. The latch is reset by the pulse at node R of Figure 3.01.1 from the receptor signal, if the state is terminated.

If the fermion's state is terminated, a new fermion is launched by the receptor emerging at the ends of the present umbilical cords. Any state change of the fermion creates a receptor, which terminates the present fermion's state and launches a new fermion. Launcher and receptor have the same logic circuit. Then that is called a receptor-launcher. When the present fermion's state is terminated, the launcher's latch is reset by the signal from node $Q^*$of the receptor, which is sent back through the lower umbilical cord. This signal is received by the input R of the launcher's latch and resets it. Then the fermion's launcher and the upper umbilical cord are wiped out. The receptor-launcher emerges automatically, when the fermion interacts with a boson. The boson changes the buffer drive current, the loading capacitor, and the buffer length of the new fermion path. If the particle interaction occurs, the fermion receptor follows the logic gate executing the interaction. The receptor-launcher transfers the capacitor, the buffer drive current, and the buffer length from the present to the future state.

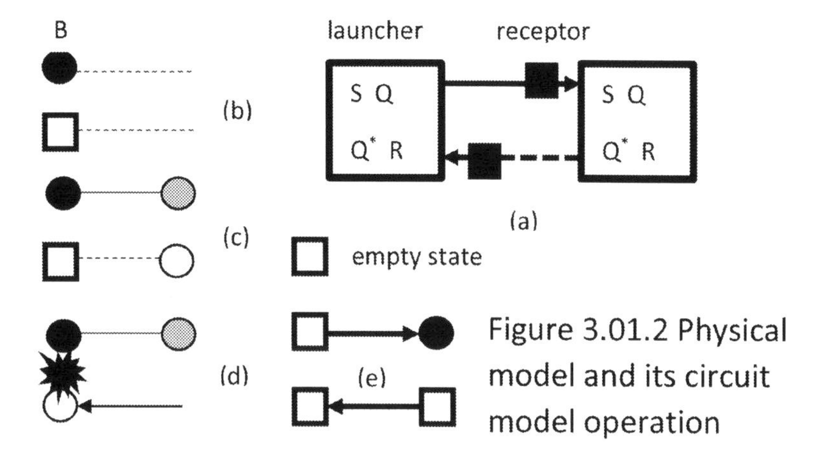

Figure 3.01.2 Physical model and its circuit model operation

How does the launcher set up the logic levels of the umbilical cords? In Figure 3.01.2(a), the launcher and the receptor circuits are same as shown in Figure 3.01.1. The launcher sends a step function, by making a LOW to HIGH logic level transition of its node Q. The step function propagates through the upper umbilical cord and reaches the receptor. The state change sets the receptor; its Q terminal sends a new fermion to the right, and its Q^* terminal sends out a virtual antifermion's step function, that is a HIGH to LOW logic level transition. This signal reaches the launcher instantly through the lower umbilical cord, and resets its latch. Then the launcher's Q terminal sends out a HIGH to LOW step signal, which reaches the receptor instantly. The receptor has been set already, so its state is unaffected, but the upper umbilical cord is wiped out.

The lower umbilical cord's logic level was HIGH before the receptor was set by the signal from the launcher. Why? When the fermion emerges at the output node of the launcher of Figure 3.01.2(b), there is no antifermion there. The fermion makes the virtual fermion and virtual antifermion pair ahead in (c). The virtual antifermion moves back and goes into a pair annihilation as in (d). This generates energy, and the antifermion vanishes. The energy goes to the virtual fermion ahead, and converts it to a fermion in (e). There is no antifermion in the lower umbilical cord. The same process repeats, and there is no antifermion in any node of the lower umbilical cord. The antifermion's empty state is logic HIGH level. Thus the lower umbilical cord is made in the process (d) and (e), such that both the input and the output nodes of the buffers of the lower umbilical cord are consistently set at the logic HIGH level. The key structure of the model of the fermion state is the upper umbilical cord, and the lower cord is made by the upper cord's activity. After termination of the present state, the lower cord's logic level is set at logic LOW level. This level is set HIGH level again, when a new fermion is launched to the state later.

The capacitors at the input node of the pulse generator of the launcher jump to the output of the launcher instantly. This is the same mechanism as the capacitor jump across the capacitor-unloaded buffer chain. In case of capacitor jump across the receptor-launcher, conversion of the step function to a narrow receptor-launcher's set-reset pulse induces the transition. This is because the step function has no place to go except for jumping across the capacitor-unloaded launcher-receptor logic circuit. When the capacitors jump, the capacitors, the buffer driver currents, and its lengths jump across the launcher's logic circuit. This jump is controlled by the conservation laws of mass (in the nonrelativistic quantum mechanics), momentum, and energy. The conservation affects the buffer drive current and the buffer length together. The pair of parameters is conserved as a single quantity (LI_0). How they are conserved is shown as follows.

Before going into the parameter jumps across the receptor-launcher, I summarize the pertinent parameter expression of the buffer chain. The common parameters are the power supply voltage V_{DD} and light velocity c (lower case) . They are the fixed model parameters. C_0 (upper case) is the load capacitor, I_0 is the buffer drive current, and L is the buffer length. I consider the case of a pair of particles joined elastically. Let the input particles' parameters have suffixes 1 and 2, and those of the exit particle have suffix 3, such that C_{01}, and C_{02} are for incident particles and C_{03} is for the exit particle. The same suffix rules follow the parameter values I_0 and the buffer length L. Using these basic parameter values, the pertinent dynamic parameter values are defined as follow

Buffer delay time = $C_0 V_{DD}/I_0$ Particle's velocity = LI_0/C_0V_{DD}
Particle's energy = $C_0V_{DD}^2/2$ Particle's mass = $C_0V_{DD}^2/2c^2$

where the last relation is the relativistic equivalent of the particle's energy and mass. If the particle's mass is m and velocity is v; the particle's momentum is given by mv and kinetic energy is $(1/2)mv^2$ in classical physics. The parameters m and v are given above.

Then, I have the following composite parameters:

Particle's momentum = $LI_0V_{DD}/2c^2$
Particle's kinetic energy = $(LI_0)^2/4c^2C_0$

The conservation laws in the simplified form (ignoring common factors) are

Mass: $C_{01}+ C_{02} = C_{03}$.
Momentum: $L_1I_{01} + L_2I_{02} = L_3I_{03}$
Kinetic energy: $[(L_1I_{01})^2/C_{01}]+[(L_2I_{02})^2/C_{02}] = [(L_3I_{03})^2/C_{03}]$

Let $L_1I_{01} = x_1$, $L_2I_{02} = x_2$, $L_3I_{03} = x_3$. Then the kinetic energy conservation is written as follows, by combining the mass conservation as

$$(x_1^2/C_{01}) + (x_2^2/C_{02}) = (x_3^2/C_{03}) = [x_3^2/(C_{01}+C_{02})]$$

and this is written, using the parameter α defined by $\alpha = C_{01}/(C_{01}+ C_{02})$ as

$$[(x_1/x_3)^2/\alpha]+ [(x_2/x_3)^2/(1-\alpha)] = 1$$

Since $x_1 + x_2 = x_3$ by the conservation of momentum, this equation can be rewritten as

$$(x_1/x_3)^2- 2\alpha(x_1/x_3)- \alpha^2 = [(x_1/x_3) – \alpha]^2 = 0$$

which has a root $(x_1/x_3) = \alpha$ and $(x_2/x_3) = 1 – \alpha$.

If this is written using the original parameters, I get the following:

$(x_1/x_3) = C_{01}/(C_{01}+C_{02})$ and $(x_2/x_3) = C_{02}/(C_{01}+C_{02})$

This is the joint conservation law of the quantity $x_1=L_1I_{01}$ and $x_2= L_2I_{02}$. L and I_0 are not separately conserved, but their product is conserved before and after the elastic collision of the particles.

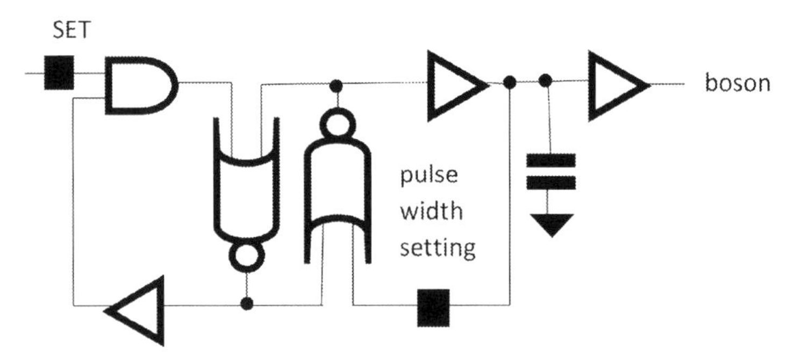

(a) boson launcher from fermion pair annihilation

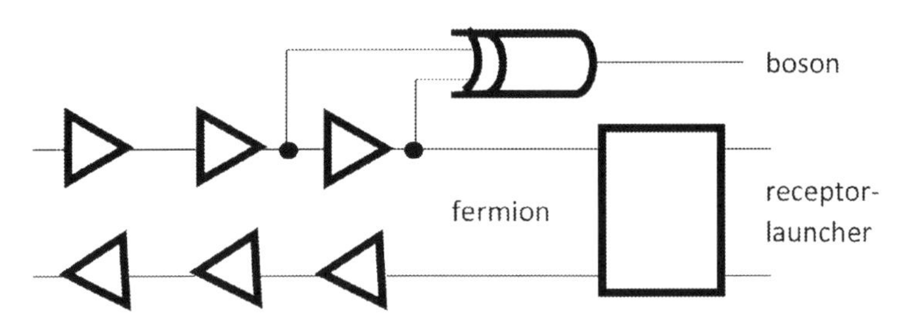

(b) boson launching from fermion

Figure 3.01.3 Boson launcher

Figure 3.01.3(a) shows a boson launcher. The core part of the launcher is the set-reset latch consisting of a pair of NOR gate loops, just as for a fermion. The boson launcher has an arrangement at its output to turn the latch off after the boson signal is sent out. The circuit determines the boson signal's pulse width t_p (section 3.03). The reset terminal of the latch is driven by the delayed boson pulse. After the reset, the latch can be set again. By repeated setting and resetting, many bosons can be packed in a single state. From this feature, a boson has no umbilical cord connecting it to the source.

A boson does not have umbilical cord if it is launched by the circuit of Figure 3.01.3(a). This type of boson launcher emerges, for instance, when a fermion-antifermion pair collide and create a boson. The boson appears, however, almost always as a pair of boson exchange configuration among the interacting fermions. The more common boson source is the fermion itself, shown in Figure 3.01.2(b). The boson exists in the space between the interacting fermions (section 3.09, in a fermion-fermion scattering). In this case, a pair of bosons makes go-and-return paths connecting the interacting fermions. This is called the *boson exchange* by fermions. The pair of boson paths is called the boson's umbilical cord.

3.02 Fermion Propagation

How does a fermion's step function propagate through the cascaded buffer chain? Let us consider the structure of the upper umbilical cord that connects the fermion source to the fermion. The upper part of Figure 3.02.1 shows the schematic of the buffer chain. Since each buffer consists of two-stage cascaded inverters, the indices of the buffer output nodes have even numbers, shown along with black circles. When a capacitor arrives at one of such nodes, a fermion is there, and it is observable. By the capacitor jump mechanism, the fermion's propagation on the buffer chain should be illustrated by the quantum mechanical model of section 2.10, by which the capacitor and charge split and move together piece by piece. Yet the illustration based on this mechanism is complicated and is confusing. So I show the illustration of the simpler semiclassical capacitor jump model. In the model's waveforms, gradually increasing voltage at the fermion's wave front is interpreted as piece-by-piece transfer of capacitor and energy, which provides the probability of the fermion's motion from a node to the next node. In the following sections, these simplified semiclassical illustrations are used for convenience.

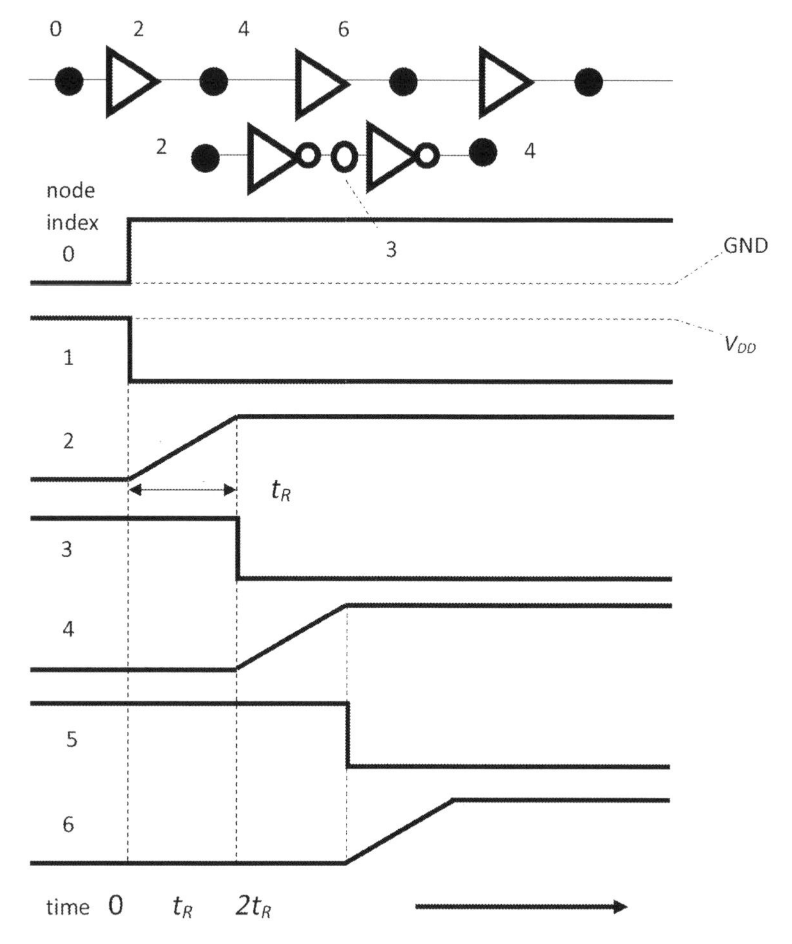

Figure 3.02.1 Fermion propagation

The lower part of the figure shows the node voltage waveforms. The horizontal axis is time, and the vertical axis is the node voltage. Node 0 makes a stepwise LOW to HIGH logic level transition at time $t = 0$. Node 0 has a charged capacitor that holds the fermion's substance. Then the output node of the first stage inverter, node 1, which was logic HIGH level before, makes instantaneous HIGH to LOW logic level transition, because the node is not capacitively loaded. Then the node 0 capacitor jumps to node 2 by the semiclassical model, and the node 2 capacitor begins to charge. Node 2 voltage goes up linearly with time, reaching the logic HIGH level V_{DD} at time t_R. Before t_R the voltage indicates the probability of the fermion to be at node 2, as I discussed in section 2.10. At time $t_{R,}$ the next inverter's output node 3, which was originally logic HIGH level, makes an instantaneous HIGH to LOW logic level transition. The node 2 capacitor jumps to node 4 and the capacitor begins to charge. The odd-numbered nodes of the chain are not capacitively loaded, and they make an instantaneous HIGH to LOW logic level transition. This waveform reshaping is required to move the wave representing the particle unchanged (*Theory of CMOS Digital Circuits and Circuit Failures*, Princeton University Press, 1992). Let the capacitor be C_0, the inverter's pull-up current be I_0, and the length of the buffer be L. Then the signal delay time per buffer stage is given by $t_R = C_0 V_{DD}/I_0$ and the fermion's velocity v is given by $v = LI_0/C_0 V_{DD}$. Here, t_R is the buffer switching delay time, not including the delay time of the signal at light velocity through the buffer's length (section 1.06).

In the fermion replacement motion described in section 2.07, the fermion at location 0 of Figure 3.02.2(a) makes a virtual fermion-antifermion pair at location 1, to move on the horizontal line H-H of Figure 3.02.2(a). The virtual fermion pair can be anywhere, such as location 2 or 3 of the figure. Then the fermion moves in the direction 0 → 2 or 0 → 3, off the horizontal direction H-H. This should not happen. The fermion at location 0 should go to location 1. For this to be possible, the fermion's state must be preprogrammed as the quantum state. The fermion state existed in the quantum vacuum, and the fermion chooses to ride on one of it.

This feature is obvious if I consider a bound state of an electron in atom. If the state into which the electron is transferred were made every time when the quantum transition occurs, it takes time for the electron to circle around the potential center to create the state. This takes time. Then the bound state must pre-exist, and the electron must ride on it, from the arbitrary entry point. The quantum vacuum is a matrix of such ready-made states in variety and in quantity. The particle's energy, the buffer drive current, direction of motion, etc., can be specified at the time of the fermion's launch.

The motion of the fermion is associated with that of the antifermion, and together they make the pair of umbilical cords (section 2.07). The antifermion can be anywhere on the cylindrical surface whose axis is the fermion's direction of motion, as shown in Figure 3.02.2(b). The loop of the associated antifermion motion can be on any plane that is hinged to the line of the direction of the fermion motion. If the fermion motion is observed along with the associated antifermion motion, the fermion's track makes a twisted cycloid-like motion. The plane on which the antifermion motion takes place is never determined by the fermion and by the initial condition.

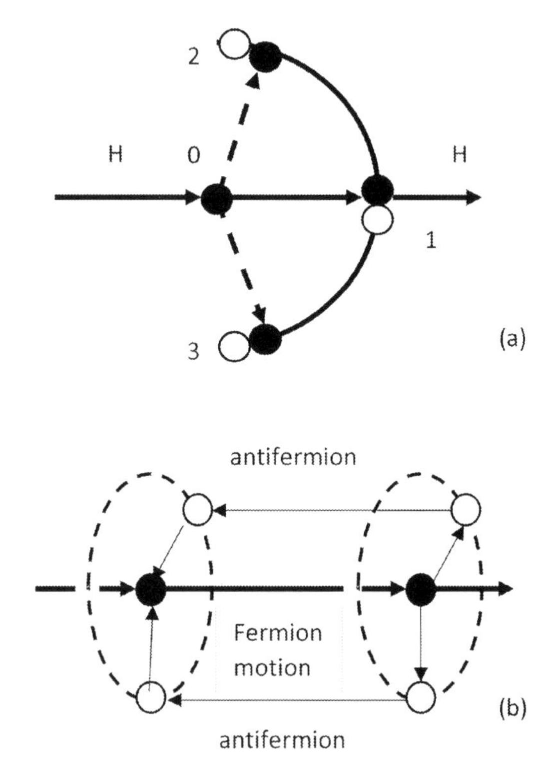

Figure 3.02.2 Structure of the quantum state

The fermion states in the quantum vacuum may have such a variety. The fermion is able to choose any such state and propagate. This variety explains many features of quantum phenomena, including the mechanism of creating its own spin (section 3.08).

3.03 Boson Propagation

Similar to the propagation of a fermion, a boson propagates in the buffer chain is as shown in Figure 3.03.1. Only the even index nodes are observable.

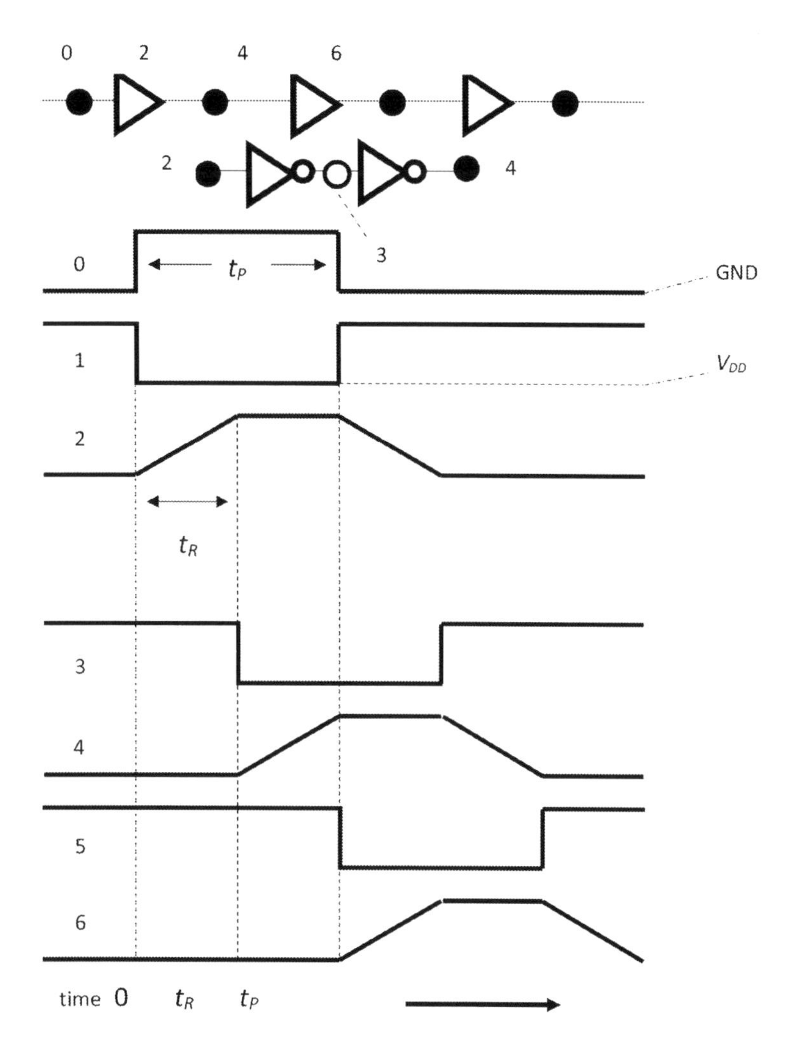

Figure 3.03.1 Boson propagation scheme

The buffer and inverter characteristics are the same as those of the fermion. A boson has one more adjustable parameter than a fermion, that is, the pulse width t_P of Figure 3.03.1. Input node 0 of the buffer chain is driven up during time from $t = 0$ to $t = t_P$, longer than the node pull-up time t_R. If $t_P < t_R$, no boson is launched to the buffer chain. If $t_P = t_R$ the boson pulse has a triangular waveform. If $t_P > t_R$, a flat top pulse is launched; t_P is the parameter that specifies the width of the boson pulse. The energy of the boson depends on the t_P/t_R ratio. If the ratio is 1, only node 2 voltage is V_{DD} at time $t = t_P$ and all the other node voltages are zero. The boson energy is $C_0 V_{DD}^2/2$. If $t_P = 2$ as shown in Figure 3.03.1, at time $t = t_P$ node 2 and 4 voltages are both V_{DD}. The boson energy is the sum of the energies of all the charged nodes, and it is given generally by $(t_p/t_R)(C_0 V_{DD}^2/2)$.

The isolated pulse on the buffer chain represents the steady propagating boson. For that purpose, a buffer must be made of a cascaded pair of inverters, and the role of the first-stage inverter is crucial. This inverter converts the slowly rising-falling waveform of the previous buffer output signal to a crisp HIGH to LOW or LOW to HIGH stepwise signal, thereby allowing it to reproduce the same waveform at the output of the following buffer. Then a narrow boson pulse is able to propagate through the buffer chain indefinitely.

By observing fermion and boson generation and propagation, I can see how the quantum space model is built from a set of functional blocks of buffers and logic gates. Each functional block specifies logic gates or buffer connection and specifies the way of the capacitor jump. They can be connected to build the path of the particle, such as the straight-line propagation, fermion-boson interaction, etc. The functional block models of the quantum states exist in the quantum vacuum in all varieties. To such a logic gate assembly, a charged capacitor representing the particle's substance *rides on*, to initiate quantum phenomena. Some gates' combination works instantly, but the input and the output signal's causal relation is always maintained. This is because the gates work in the *minimum unit of time*, which is regarded as zero, yet the time's order secures the causal relation (section 1.06). Causality is kept under all circumstances. Time is continuous like real numbers, but the model's instantaneous operation maintains the order. This feature exists clearly in the digital equivalent circuit model operation in the idealized limit of space-time. The *race condition* of the logic circuit is avoided in the model.

The fermion model I described before has only one charged capacitor at the end of the buffer chain. This model can be generalized such that the several nodes behind the front are capacitively loaded. Each capacitor simultaneously jumps forward as the fermion moves, and maintains the structure. The basic function of such a fermion is the same as a simple fermion, but the fermion's energy is higher than that of a single capacitor. Such a fermion's waveform looks like boson, having long t_P. This similarity has a consequence, when many fermions make a composite particle. The nucleus of U^{235} is a fermion, and that of U^{238} is a boson. As atomic nuclei, the two are basically quite similar. The model covers such cases.

3.04 Fermion-Boson Interaction

Nuclear interaction is emission or absorption of a boson by a fermion. Its equivalent circuit model is shown in Figure 3.04.1(a) and (b), respectively. In Figure 3.04.1(a), the fermion emits a boson. The boson is created by using the delay time of the fermion propagation path and an exclusive OR gate. As discussed in section 3.01, this is the common boson generation mechanism of fermion-fermion interaction. Figure 3.04.1(b) shows the fermion absorbing a boson by fermion-boson fusion. Both cases involve logic gates to execute the interaction. The logic gates that execute the processes are exclusive OR and OR gate. The two types of processes may occur as a pair, at the time when two fermions interact (section 3.09). Bosons are *exchanged* between the interacting fermions to exercise force.

Because the process is subject to energy and momentum conservation of the particles, the analysis is more easily described by using conventional classical mechanics parameters. So in boson emission, the incident fermion has mass m_F and velocity v_F, the emitted or absorbed boson has mass m_B and velocity v_B, and the exit fermion has mass m_{FE} and v_{FE}, respectively. To make the analysis simple, the incident fermion, the emitted or absorbed boson, and the exiting fermion are all on a single line.

Figure 3.04.2(a) shows the momentum vector diagram of emission of a boson from a fermion and Figure 3.04.2(b) that of absorption of a boson. Since nuclear reactions are inelastic, it is necessary to show that energy is converted between the kinetic and the *internal* energy of the particle. The kinetic energy in classical particle collision is converted to heat or internal elastic energy of the exit particle. This is not the case in elementary particles.

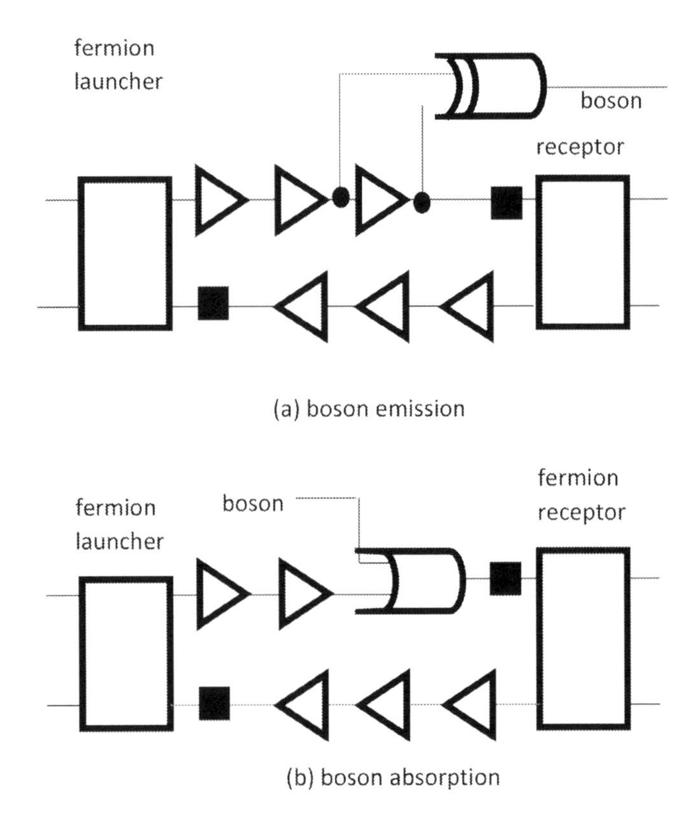

Figure 3.04.1 Fermion-boson interaction

Elementary particles have no internal structure or material to be heated, so the particle's mass changes. Then the exit fermion's mass m_{FE} is affected. That is, m_{FE} is not equal to $m_F - m_B$ (boson emission) or $m_F + m_B$ (boson absorption). In case of boson emission, I try to find the correction to the approximated mass, $m_F - m_B$. Let us consider the boson emission by a fermion of Figure 3.04.2(a). If the boson energy is small, the exit fermion mass m_{FE} must be close to $m_F - m_B$. I try to find the highest order correction to this approximate mass δm_{FE}, defined by $m_{FE} = m_F - m_B + \delta m_{FE}$. The analysis involves relativistic mass energy, which is proportional to c^2, where c is the light velocity, that is, a large number. I execute the algebra in c^0 order in the momentum conservation, and derive the highest order correction of the mass of the exit fermion, that is of the c^{-2} order, assuming that the emitted boson energy is small. The momentum conservation is written as

$m_F v_F = m_B v_B + m_{FE} v_{FE}$ or $v_{FE} = (m_F v_F - m_B v_B)/m_{FE}$ (1)

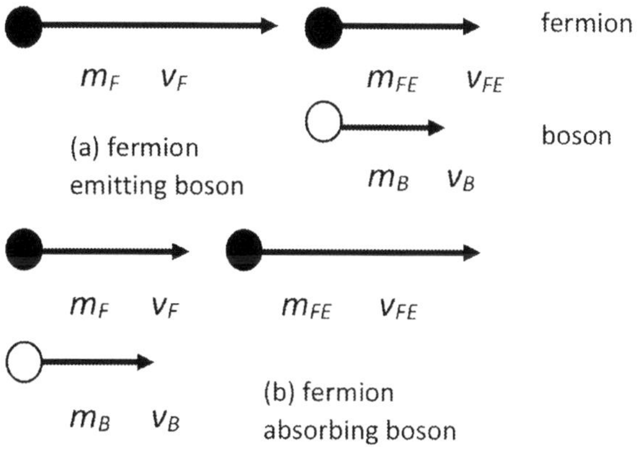

fermion

m_F v_F m_{FE} v_{FE}

boson

(a) fermion emitting boson

m_B v_B

m_F v_F m_{FE} v_{FE}

m_B v_B

(b) fermion absorbing boson

Figure 3.04.2 Definition of particle interaction parameters

I write the energies of the incil write the energies of the incident fermion E_F and that of the emitted boson E_B including the mass energy as

$E_F = m_F c^2 + (1/2)\, m_F v_F^2$ and $E_B = m_B c^2 + (1/2)\, m_B v_B^2$ (2).

The formula includes the first order correction of the particle's relativistic mass energy by the particles' kinetic energy gain or loss. Correction of the total energy is found from the energy conservation relation

$E_F = E_B + m_{FE} c^2 + (1/2) m_{FE} v_{FE}^2$ (3).

By eliminating v_{FE} by using Eq. (1), I get the equation satisfied by m_{FE} as

$$c^2m_{FE}{}^2 - (E_F - E_B)m_{FE} + (1/2)(m_Fv_F - m_Bv_B)^2 = 0 \ (4)$$

I seek for the most significant correction term of m_{FE} to the approximate mass m_{FE}, that is $m_F - m_B$. This is the compatible degree of approximation to E_F and E_B expression, assuming that light velocity c is high, and the term including $1/c^2$ is small. By expanding the solution of Eq. (4) in series of $1/c^2$ and choosing the physically meaningful solution that is close to $m_F - m_B$, I get

$$\delta m_{FE} = - [m_Fm_B(v_F - v_B)^2]/[2c^2(m_F - m_B)] < 0 \ (5)$$

This is the correction to the mass $m_F - m_B$. Since $m_F >> m_B$, the fermion mass decreases. That is, some part of the fermion's mass energy is carried away as the kinetic energy of the boson. This expression is rational, since $\delta m_{FE} = 0$ if $v_F = v_B$. Then the incident fermion simply splits into two pieces and all the particles move by the same velocity. If m_B is close to m_F, the boson takes most of the energy away from the fermion, and the approximation fails.

In the second case of boson absorption, the momentum conservation equation is

$$m_Fv_F + m_Bv_B = m_{FE}v_{FE} \ (6)$$

and again executing the algebra in the c^0 order terms in the momentum conservation, the total energy conservation is written as

$$E_F + E_B = m_{FE}c^2 + (1/2)m_{FE}v_{FE}{}^2 \ (7)$$

and E_F and F_B are same as Eq. (2). The equation satisfied by m_{FE} is

$$c^2m_{FE}{}^2 - (E_F + E_B)m_{FE} + (1/2)(m_Fv_F + m_BV_B) = 0$$

I seek for the solution of this equation which is close to $m_{FE} = m_F + m_B$. The mass correction is defined by $\delta m_{FE} = m_{FE} - (m_F + m_B)$. Then δm_{FE} is

$$\delta m_{FE} = [m_Fm_b(v_F - v_B)^2]/[2c^2(m_F - m_B)] > 0$$

This means that the mass of the exit fermion increases by acquiring the boson's kinetic energy. This formula is also rational since $\delta m_{FE} = 0$ if $v_f = v_B$. The inelastic interaction of fermion and boson changes the fermion's mass.

3.05 Wave and Particle Model

I described the de Broglie wave model of the linear resistance-capacitance ladder circuit in sections 1.05 and 2.01, and the particle propagation model of the capacitor-loaded digital buffer chain circuit in sections 2.09 and 2.10. How are the two models mutually related to describe the particle-wave double features of quantum phenomena? The linear de Broglie waves are assembled to a wave packet, and the motion of the wave packet as a classical particle is described by the digital circuit model. Yet the de Broglie wave was originally defined on the particle making the classical motion. The two models are mutually dependent. Their relation must be clarified.

A particle is a localized structure, and a wave is an extended structure in space. How can the two different spatial structures be made compatible? This is a basic question of quantum mechanics that seems to conflict with the common sense of classical physics but is not properly addressed in standard textbooks. I examine this feature and the related issues in this section. I begin with the digital equivalent circuit model.

Let the particle's propagation path parameters be capacitance C_0, buffer drive current I_0, buffer length L, and power supply voltage V_{DD} (section 3.01) as before. Then the mass of the particle m is given by $m = C_0 V_{DD}^2/2c^2$, where the lowercase c is the light velocity. This is the relativistic *equivalence* of energy and mass. The classical particle's velocity v is given by $v = (LI_0)/C_0 V_{DD}$. The classical particle's momentum p is given by $p = mv = V_{DD}(LI_0)/2c^2$. Then the de Broglie wave length λ is given *by* $\lambda = 2\pi\hbar/p = 4\pi\hbar c^2/V_{DD}(LI_0)$, and the wave number k is defined by $k = 2\pi/\lambda = V_{DD}(LI_0)/2c^2\hbar$.

This buffer chain models the classical particle motion, which leaves several conflicting features of classical and quantum mechanics. Specifically, in the model, the particle exists at the location where the capacitor exists;

that is, the location of the particle x appears definite. The particle also carries definite momentum p. The two definite parameters violate the uncertainty principle, that their measurement uncertainties must satisfy $\delta x \delta p = \hbar/2$. How is this conflict resolved?

Each buffer of the buffer chain of section 3.02 has definite length L. The momentum $p = V_{DD}(LI_0)/2c^2$ is proportional to the buffer length L. In the proper quantum model, both the particle's location and momentum must have uncertainty. Then not one but many buffer chains must be superposed probabilistically, and each of which must have a different buffer length, as shown in Figure 3.05.1(a)–(c). The buffer chain that gives the most significant feature to build up the wave packet has the buffer length L shown in Figure 3.05.1(c). The others provide its harmonics. The superposed set of buffer chains gives the proper uncertainty of the particle's momentum. The particle's momentum given before is the measure of the momentum uncertainty. The most significant buffer length L gives the measure of the uncertainty of the particle's location. The particle is somewhere between the two consecutive buffers having length L of the chain.

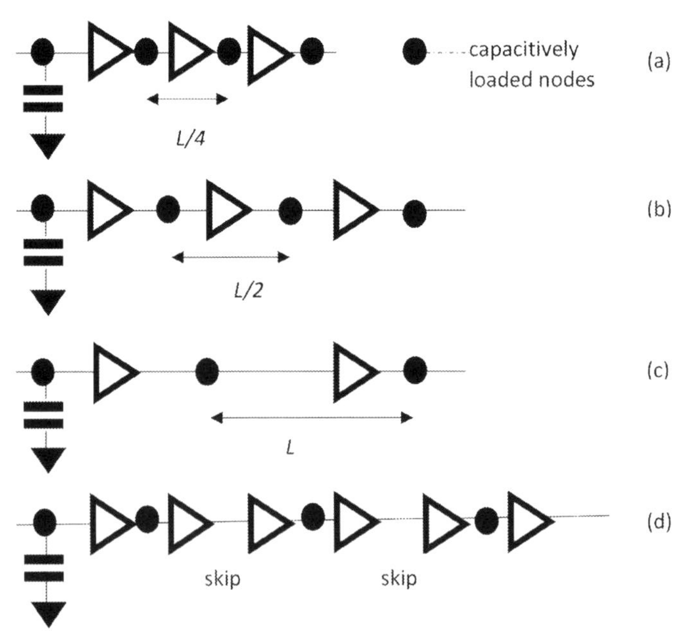

Figure 3.05.1 Buffer length and capacitor skipping

The length of the buffer L should be about the half the wavelength of the most significant de Broglie wave's component creating the peak of the wave packet representing the particle. From the two uncertainties, I can get the equivalent of the uncertainty formula of the buffer chain model as follows. Let the most significant wavelength be λ, and p_x has uncertainty

$$\delta x \delta p_x = (\lambda/2)\delta p_x = [2\pi \hbar c^2/V_{DD}(LI_0)][V_{DD}(LI_0)/2c^2] = \pi\hbar$$

This is a formula like the uncertainty principle. In this formula, $\delta p_x = V_{DD}(LI_0)/2c^2$ means to cover from 0 to ∞ of the momentum range to give the uncertain de Broglie wave length. This is certainly an overestimate of the wave components to build up a wave packet. By moderately estimating the required number of harmonics of the de Broglie wave to build the wave packet is up to 7th harmonics, the right side of the formula must be divided by 7, or approximately by 2π. Then I get the conventional uncertainty relation of the equivalent circuit model, that is $\delta x \delta p = \hbar/2$.

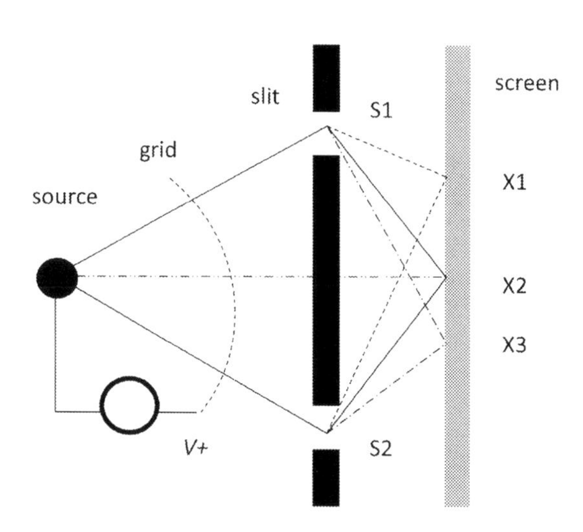

Figure 3.05.2 Electron beam interference

In this discussion, I assumed that the range of uncertainty of the momentum is just wide enough to build a nonzero width wave packet. In a charged particle like an electron, it is possible to create a particle beam that has precisely-defined velocity and momentum by certain test arrangements. Then, the above limitation does not apply. One realistic case is the double slit interference experiment shown in Figure 3.05.2. An electron can be accelerated to a definite velocity if voltage V is applied to the accelerating grid. Then the electron's kinetic energy will be eV, where e is the electronic charge. This energy is the kinetic energy of the electron, that is $(1/2)mv^2$ where v is the electron velocity and m is its mass. Then, the electron's classical momentum is given by $p = (2meV)^{1/2}$ and the de Broglie wavelength $\lambda = 2\pi\hbar/(2meV)^{1/2}$. There is no uncertainty of the momentum of the electron in this experiment.

The electron's de Broglie waves passing through the slits S1 and S2 of Figure 3.05.2 can interfere constructively, if the path lengths from the grid, via slits S1 or S2 and the destination on the screen X differ by an integral multiple of the de Broglie wavelength λ. At such location, the probability of detecting the electron peaks. If the path lengths are different by an odd number times the half wavelength $\lambda/2$, the probability of finding an electron there is zero. The de Broglie wave model is possible because the electron can be anywhere as momentum uncertainty is zero.

Let us consider the center location X2. The two paths from the grid via slits S1 and S2 to X2 have equal length; here I expect a constructive interference. How about location X1 or X3, where the electron is also detected? The path length through slit S1 to X1 is shorter than the length through slit S2 to X1. If the electron is a de Broglie wave having fixed momentum and wavelength, this is expected, because the electron is spread out over the entire path by the uncertainty principle. If electron is considered as a particle, it probabilistically splits into a pair of paths, and each path goes through independently through each slit S1 and S2. The *partial* electron via slit S2 must move faster through the buffer chain than the partial electron via slit S1 to catch up and *integrate* to a whole electron at location X1. Is this possible? If not, does the digital particle model fails to explain this effect?

The conventional explanation of electron diffraction is that the electron makes a de Broglie wave, which interferes. Yet, the de Broglie wave itself is a part of the quantum mystery. I try to supplement the wave explanation by the particle propagation model to make it plausible. Suppose that a single electron left the grid and is detected at location X1 of Figure 3.05.2. Then isn't it natural to think that the *imaginary* split electron pair goes through slit S1 and S2 separately, and then meets and unites at location X1 as a whole electron? In my digital equivalent circuit model, this explanation is made possible by the model's special feature. Skipping the capacitor jump is the mechanism to adjust the electron's path length or its velocity. Figure 3.05.1(d) shows my interpretation. The figure shows a capacitor jump mechanism to make the particle move fast through the buffer chain by skipping some buffers. Let us ask a question, if the path difference is as it is observed by physicists? Or as it is *sensed* by the particle pair itself?

A buffer chain can be compacted. A capacitor may jump across the buffer nodes of the compacted buffers of path S2X1 of Figure 3.05.2. A *compacted* buffer has the same drive current, the same length, and the same load capacitor as *each of* the skipped buffers. If buffers are skipped, the distance sensed by the particle gets effectively shorter, or the velocity of particle sensed by it effectively increases. The S2X1 partial electron *senses* as if the path length were shortened to become the same as that of the S1X1 partial electron. The longer path of Figure 3.05.2 takes this option at some part of the path to catch up the shorter path. The de Broglie wavelength remains the same, but the path length of the electron effectively decreases. This mechanism also explains why a fixed-momentum particle can be anywhere in its path by the uncertainty principle. The time and distance in the interference experiment are not those of the observing physicists, but those sensed by the interfering particle itself. I explain this feature in the digression of the next section.

When the particle rides on the existing quantum state, it does not necessarily propagate from a buffer to the next buffer as shown in Figure 3.05.1(a) to (c). Some buffers are skipped as shown in Figure 3.05.1(d). Which buffer is skipped depends on chance. In Figure 3.05.2, let the length of the buffer be L and let the distances from the grid to slits S1 and S2 be the same. If constructive interference occurs at location X1, distance S2X1 is longer than distance S1X1 and each path contains $b2$ = S2X1/L buffers and $b1$ = S1X1/L buffers, respectively. Then, if $b2$-$b1$ buffers of the path S2X1 are skipped, the number of buffers of the both paths become the same. At the location of the constructive interference, this buffer number adjustment takes place, and the split electrons reach the destination at the same time.

I seek the relation between the digital and linear equivalent circuit model further. Let us assume that the particle does not skip buffers. Figures 3.05.3(a) and (b) show schematically the voltage waveforms of the digital and the linear circuit models, respectively. The de Broglie wave packet represents the particle's existence probability. The particle itself is practically sizeless, but the wave packet has nonzero width, because of the uncertainty between the particle's location and its momentum by the uncertainty principle. A fermion's digital waveform is a step function, and the particle exists at the edge of the step. The switching waveform of the semiclassical model shows a linear increase of the node voltage with time. In the model. I interpreted the linear increase of the node voltage as the *effective node voltage*, that is the measure of the probability of the fermion at node N1 and N3. In the quantum digital circuit model, the wave packet's probability distribution is created by the step-by-step capacitor jump from node N1 to node N3 as follows. At the start, the fermion is 100 percent at location N1.

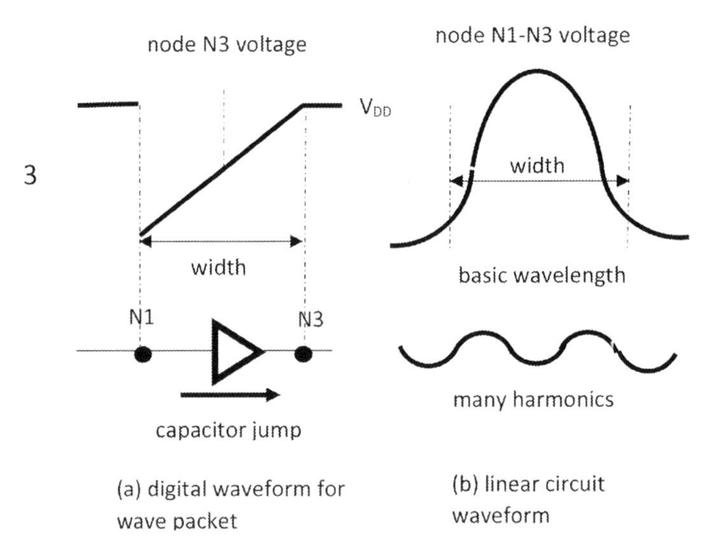

Figure 3.05.3 Wave packet and particle motion correlation

As the capacitor and charge are sent to node N3 step by step, The probability of the fermion being at location N1 decreases, and the probability of its being at location N3 increases. N1-N3 distance is the particle location uncertainty. When all the capacitor and energy are transferred from node N1 to node N3, he fermion is at node N3 100 percent. A fermion exists only if there is full capacitance C_0 and its terminal voltage is V_{DD}. As the particle arrives at node N3, the next step, trying to move the fermion further to the right, begins. Then, the probability of fermion at node N3 decreases, and makes the bell-shaped probability distribution that simulates the wave packet's probability profile emerge. By comparing the effective node voltage and the probability given by the wave packet, the length of the buffer L must be set equal to the width of the wave packet. This is a rational approximation of the uncertainty principle; the particle is assured to be in the range between the pair of consecutive buffer nodes. From this, the circuit-theoretical feature matches well with the uncertainty principle of quantum mechanics. That is, how linear and digital circuits are mutually related. The digital circuit is really a linear circuit that saturates in the HIGH and LOW power supply voltage levels.

Conversely, the linear wave model is the limit that the digital transient region of Figure 3.05.3(a) extends out over the entire domain of the particle motion. The digital circuit works as a linear circuit only during the switching transient. That is how the linear and the digital circuit models are related. Thus, the digital circuit model is also proper to describe the effect. In a confined quantum state, the digital circuit's transition region spreads out over the entire particle's dynamic region. This is relevant to the particle bound by a potential well, as I discuss in the next section.

3.06 Relation between Linear and Digital Circuit Models

The quantum states that occupy the digital circuit model's switching transient region display quantum phenomena bound in a narrow space. Figure 3.06.1(a) shows the digital circuit model and Figure 3.06.1(b) the linear circuit model of states having such limited dynamic space. The digital circuit model shows a potential well that is still wide enough, having a flat bottom. The particle between the hard potential walls is reflected by them, but between them it is free. The walls work as a pair of receptor-launchers. The receptor-launcher switches alternately to execute the particle's *catch-ball* between the walls, thereby creating a standing wave of the particle having certain wavelength. A standing wave state has a definite energy.

Figure 3.06.1(b) shows that the potential well is as narrow as only one buffer length of the digital model. It allows only an integral number of de Broglie waves fitted in the narrow space. The figure shows the linear equivalent circuit model supported by a unity gain linear buffer. In this linear de Broglie wave propagation model, the capacitors create the model of the probability wave's propagation path with the imaginary value resistor (section 2.01). An example is the low-energy bound state of an electron in an atom. Electrons are bound in the potential well created by the positively charged nucleus at the center of the atom. Each of the electron's confined de Broglie waves satisfies the boundary condition set by the potential well. Each standing wave state has its own discrete energy. The electron makes transition from one state to the other state, by emitting or absorbing a photon, a boson. Figure 3.06.2(a) shows the electron's energy levels in an atom. The potential made by an electrically positive nucleus is spherically symmetric, and the potential's profile looks like a funnel, whose upper opening extends to infinity. The lowest energy ground state has only one de Broglie wave period fitted in the two-dimensional ring orbit.

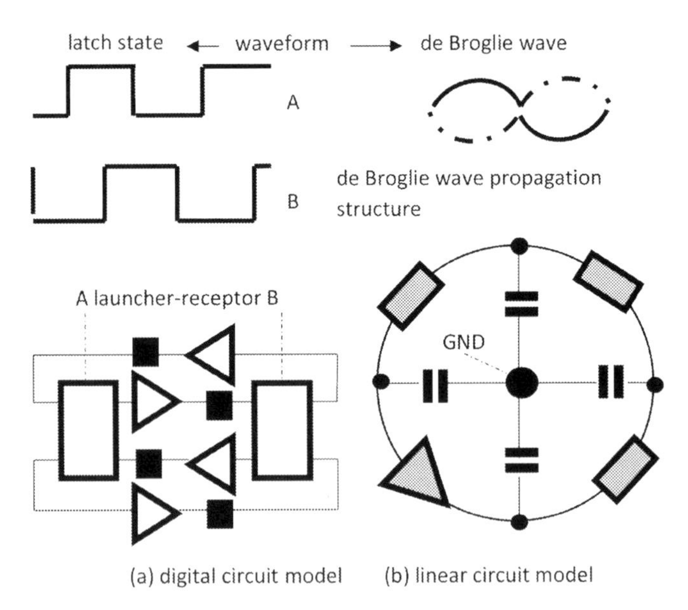

Figure 3.06.1 Bound state of particles

As the electron energy increases, the number of the de Broglie wave periods accommodated by the circular orbit around the nucleus increases one by one. There are countably many quantum states above the *ground state*, because the funnel of the electric field potential extends to infinity. The energy levels of the states are discrete, but as the energy level gets higher, the energy gap between the successive energy levels becomes smaller, and finally becomes almost continuous. In such a state, the electron is partially free from the potential source, and the energy levels become almost continuous, like free electron. At this free energy boundary, the number of de Broglie waves increases from countably many to continuously many. This is because the range of the particle's activity extends from the limited to the entire free space.

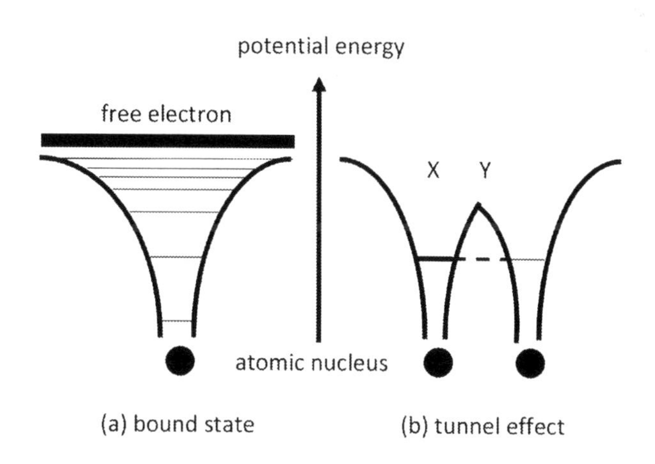

Figure 3.06.2 Structure of bound state

From this observation, at the atom's low energy levels, the quantum structure cannot be modeled by the digital equivalent circuit, since the wave packet cannot be built by superposing only a small number of de Broglie waves. The particle image fails. For the low energy levels confined within the narrow potential well, the proper model is the standing de Broglie wave fitted in the orbit, described by the linear circuit model. There can be several to countably many wave periods, each of them having a specific energy. If the bound energy level increases, the number of de Broglie wave periods accommodated in the higher energy state increases, and the energy gap between consecutively increasing energy levels becomes smaller, and the consecutive states get similar. The waves in the higher states are able to be superposed to make a structure like a wave packet.

As the energy of the confined electron gets higher, the energy-wise closely spaced quantum states can be superposed, and the electron's wave simulates the wave packet circulating around the nucleus. Finally, just before the electron becomes free, the electron motion becomes similar to the circular rotation of the wave packet around the central nucleus. The state becomes similar to the classical electron motion around the nucleus like that of a planet in the solar system. I showed how the confined wave of a bound electron changes continuously to the state of a free electron.

The transition from wave to particle is gradual and continuous. But the digital circuit model reveals a character shared by both bound and free particles, that is, the difference between a fermion and a boson. The digital circuit model has a feature qualitatively not contained in the linear circuit model. The fermion-boson difference is the basic feature of the digital equivalent circuit model of quantum phenomena. In the linear circuit model, the fermion-boson difference is set up mathematically by the symmetrized and anti-symmetrized wave functions. This appears artificial. The linear equivalent circuit model is a quantitative model that develops all the times of the system, whose structure do not change with time. The model is *analytical* in the time variable. The state develops continuously in time, once the initial condition is set. Introducing complex parameters and probabilistic interpretations in the quantum mechanics does not change this feature.

Besides this feature, the digital equivalent circuit model distinguishes the past, present, and future clearly. The fermion and boson emerge from this difference. The digital circuit model does not predict the indefinitely remote future. For instance, generation or annihilation of a particle is not covered by the linear circuit model. In the digital circuit model, generation and annihilation of the particle can be handled simply by a properly modeled logic circuit. Mathematically, the digital circuit is a self-altering circuit, and because of that, the time cannot be analytic. The difference between the linear and the digital circuit model means a difference in the model's mathematical nature.

This is significant; because the time need not be an analytic parameter in the digital circuit, the model's structure is free to change with time.

Another feature of the bound state occurs when the potential wells are spatially close nearby, such as that shown in Figure 3.06.2(b). Because of the probabilistic nature of the de Broglie wave, the wave penetrates into the barrier between the two potential wells. Then the wave function extends from one well to the next well as shown in Figure 3.06.2(b). The probability depends on the barrier's height and thickness between the potential wells X and Y. If the barrier is low, the electron that existed in well X *leaks* into well Y. The particle that existed in well X may be found in well Y later. This effect is called the *tunnel effect.* The tunnel effect is conventionally explained by the wave model. I suspect that if this effect could also be described by the particle model, a particle in a potential well can be temporarily energized by the quantum vacuum fluctuation and gain enough energy to go over the potential barrier.

The tunnel effect is usually explained using the structure that has a potential barrier, as shown in Figure 3.06.2(b). The case of skipping the capacitor jump in the last section is a special case of the tunnel effect. Since there is no potential barrier, the effect occurs freely along the particle's entire path. This explains why an electron having a fixed momentum propagating along a straight line can be anywhere on the path. This explanation makes the de Broglie wave less mysterious. This explains also what is physically meant by the uncertainty principle. The electron at any location repeats tunneling to the next location in the propagation direction, and as a consequence, it makes a spatially spread-out wave-like structure. In the experimental arrangement of Figure 3.05.2, the electron has fixed momentum, so it can be anywhere in the space between the shadow and the screen. Then it is able to interfere. This is an easy way of visualizing the de Broglie wave, which never had an intuitive visual model; the barrier-less tunnel effect is the mechanism of de Broglie waves and of the uncertainty principle, and its consequence is skipping the capacitor jump in the digital equivalent circuit model. Thinking in this way, a new insight into quantum mechanics emerges.

Digression: Who Observes What?

Standard textbooks of quantum mechanics introduce de Broglie waves at the beginning, by showing the particle diffraction experiments. From this explanation, the reason why the de Broglie wave is basic to quantum mechanics is not obvious to the beginners. The wave remains a mystery. Isn't it pedagogically better to begin with the particle image—that is, to start from the uncertainty principle? The principle at least has a physically clear image; it is explained by the determination of the particle's position by an optical microscope that illuminates the particle by light having wavelength λ to determine its location with the accuracy of λ. Since the photon has momentum $p = 2\pi\hbar/\lambda$, the momentum of the particle reflecting the photon gets that much ambiguity. Then, the location of the particle and its momentum get a complementary relationship that affects the spatial structure of the particle's trajectory. Figure 3.06.3 shows the uncertainty of the spatial position and its momentum as *observed from the particle*. Observation by the particle is more proper to determine its interference.

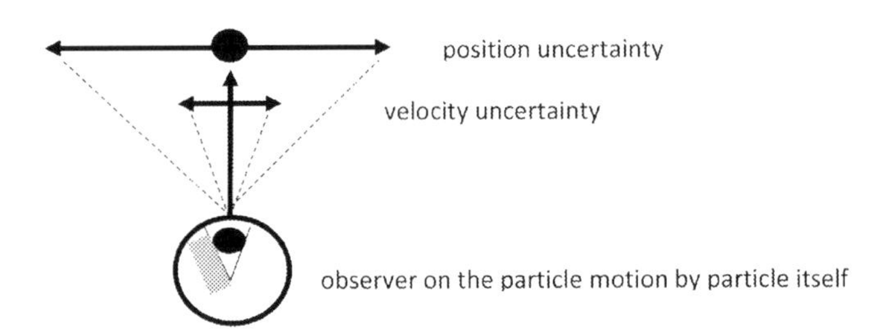

Figure 3.06.3 Length of path as observed by the particle

Then the path length difference of the interfering particle of Figure 3.05.2 can be explained not by the difference of the spatial distance measured by the outside observer, but by the distance *sensed by the pair of partial particles themselves*, which feel the same distance. The two distance measures are different. This is the key point; then it becomes possible for the partial particle pair to go through the separate paths and then unite at the same location at the same time. The pair integration by themselves by their own sensing of space and time appears more natural than introducing the probability wave interference. This explanation is compatible with the buffer skipping propagation of the digital circuit model of section 3.05.

3.07 Spin and Equivalent Circuit Waveforms

Spin is the angular momentum of a small, ideally a point particle. Its classical model is a ball rotating around its symmetry axis. Can a size-less elementary particle have angular momentum? If it does, it should also depend on the particle's mass. Yet the heaviest lepton (tauon) and the lightest neutrino (electron neutrino) have the same spin, $\hbar/2$. Spin appears to be independent from the particle's mass. Then, spin could be independent from the particle ball's rotation, and it is a more basic property of the particle than its mass. The mechanism of an elementary particle's spin has been the most difficult feature for me to understand. I saw that no textbook explained how to think of it, beyond giving the relativistic mathematics. Then I had to think its model by myself. To get insight into the spin, I first consider the relation between the equivalent circuit waveforms and the spin's magnitude.

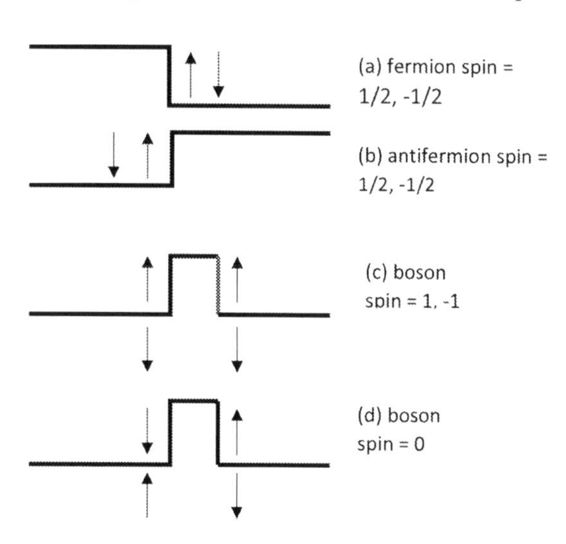

Figure 3.07.1 Spin and transition edge

The fermion and the boson are represented by step function and isolated pulse, respectively, in the equivalent circuit model. Spin of the elementary fermion comes in the basic unit of $\pm\hbar/2$, and it must be associated with the single transition edge of the step-function waveform of the fermion or antifermion, as shown in Figure 3.07.1(a) and (b), respectively. A boson, having a pair of transition edges of an isolated pulse, appears to carry spin $\pm\hbar/2$ each, as shown in Figure 3.07.1(c) and (d). Spin polarity at each transition edge is either up or down, independently. Spin assignment at the waveform's transition edge suggests why an elementary fermion has spin $\pm\hbar/2$, and an elementary boson has spin $\pm\hbar$ or 0. The fermion and boson seem to generate angular momenta at the waveform's transition edge or edges independently, and they add up. The particle's substance must be rotating at the edge of the waveform of the particle's equivalent circuit model. I must go back to the physical model on which the equivalent circuit model was built. Let us consider this possibility in the next section.

Is there any hint to this speculation? A ball has two angular degrees of freedom; the two angles set the direction of the axis of rotation. Rotation of the ball around the axis determines the magnitude of angular momentum. By going from a nonzero size ball to a point, where do the two degrees of freedom go? In the limit of minimum space-time domain, where no object has any definite shape, the angular degree of freedom could change its character. Especially, could the axis of rotation go through the point particle in this shape-less world? As a possibility, the point particle may not rotate around its axis, but the particle itself may rotate around the axis outside of the particle, I suspect. I explore this possibility, if it has any merit or not.

If I look at this spin assignment at the transition edge or edges of the waveform, it appears not possible, or at least not natural, to create a boson companion of a fermion and vice versa. From this graphical model, supersymmetry appears, at least, awkward. In spite of intensive search since the 1990s, not a single *super partner* has been discovered by this time.

3.08 Model of Spin

I suggested that fermion and boson particles are making *some sort* of rotation at the location of the transition edge of the model equivalent circuit's waveform. If this is the case, I am able to explain the spin angular momentum by the angular momentum quantization rule of early quantum mechanics used to derive the spectrum of the hydrogen atom by Bohr. According to this mechanism, spin is a necessary by-product of the particle replacement motion of fermion and boson, discussed in section 2.07 and 2.08, respectively. This explains a few other features of the elementary particle's spin.

Figure 3.08.1 shows fermion propagation by the particle replacement mechanism. The fermion shown by the black circle makes a virtual fermion shown by the gray circle in the direction of motion. This virtual fermion is created there from the virtual boson, by emitting the virtual antifermion by the process 1, which is shown above and below the virtual fermion, depending on the spin's direction. The spin's direction follows the right-hand rule ("up" means from the page to you). The virtual antifermion (white circle) moves back to the location of the original fermion by the process 2, and then fuses with it by process 3, thereby letting the fermion emit its energy. The virtual fermion-antifermion pair annihilates. The released energy travels to the virtual fermion to the right in process 4, and converts it to a fermion. The virtual antifermion and the energy motion 1→4 closes a loop as the fermion makes one step of motion to the right. Here, I made an assumption: a closed loop need not be made by a single particle, such as an electron in Bohr's hydrogen atom; the virtual antifermion motion can be supplemented by the motion of the fermion's energy motion (process 4) to close the loop.

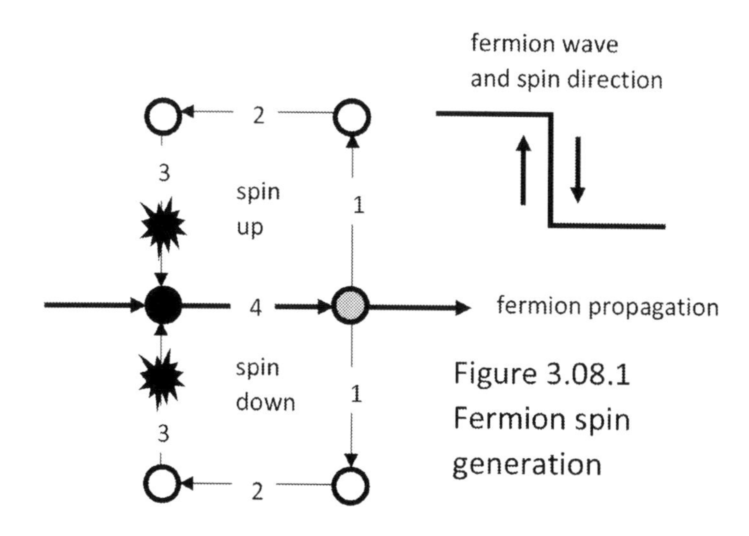

Figure 3.08.1
Fermion spin
generation

It is natural that a fermion, which has mass, creates angular momentum by rotation. But the virtual antifermion carries no mass. Then why does its loop motion make angular momentum? Here is an interesting point. Bohr's angular momentum quantization rule, written in the classical mechanical notation is, *angular momentum* = $mr^2\omega = \hbar$ for the hydrogen atom's ground state, where r is the radius of the electron orbit and ω is the rotation's angular frequency. The electron may have whatever the mass m or r, to create angular momentum $\hbar$. So whatever mass the rotating particle may have, the angular momentum of the closed loop motion has the fixed angular momentum $\hbar$ in the ground state. Then the spin angular momentum is independent from the particle's mass, as we see in nature (the heaviest lepton and lightest neutrino have the same spin).

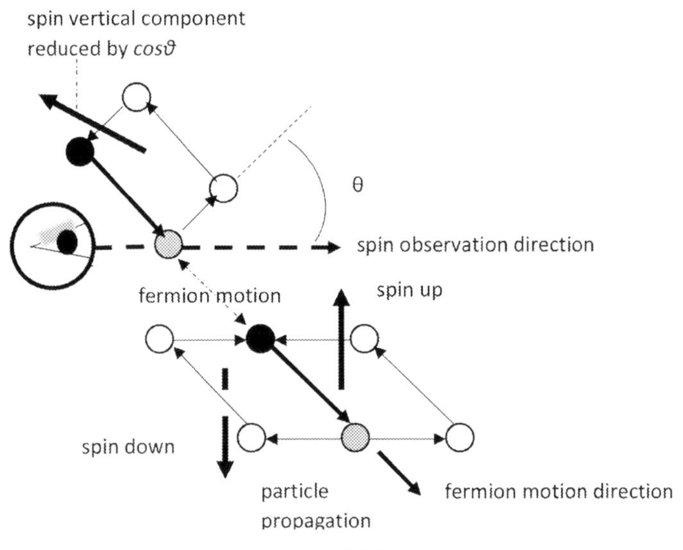

Figure 3.08.2 Fermion spin $\hbar/2$ formation

Yet there is a discrepancy of factor 2 to the spin angular momentum. That is due to the angle of the plane, on which the antifermion motion takes place. The plane may have any angle if observed from the direction of the fermion's motion. The loop of the antifermion rotation is on the plane *hinged* to the line of motion of the fermion, but it can make any angle. Looking from the direction of motion of the fermion, the angle that the plane can make has 360 degrees freedom. This angular freedom explains the factor 2 on average.

I show the angle of the plane, ϑ, on which the rotational motion of the virtual antifermion takes place, in Figure 3.08.2 (above). The angle of the plane of the virtual antifermion motion seen from the direction of the spin observation is ϑ in Figure 3.08.2. The associated virtual antifermion's motion can be on any plane making any angle relative to the horizontal direction of the observation, which is perpendicular to the direction of motion of the fermion.

Since the angle ϑ is arbitrary, the motion of the fermion and the associated virtual antifermion put together looks like a twisted cycloid, as shown in Figure 3.08.3. The center line of the cycloid is the path of motion of the fermion. Only this center line, the fermion's path, is externally observable. The planes of the associated virtual antifermion can make any angle hinged to the fermion's propagation path, so the *cycloid* freely *snakes around* the path of the fermion path, as shown in Figure 3.08.3. The angle ϑ is random.

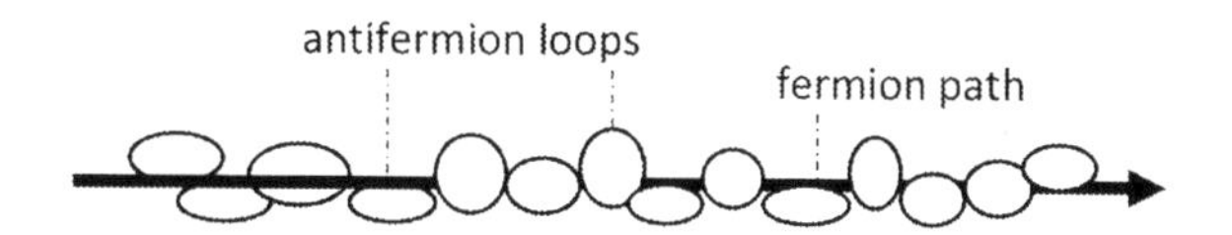

Figure 3.08.3 Twisted cycloidal path of fermion

According to Bohr's quantization rule of the ground state electron of a hydrogen atom, the angular momentum of the circulating electron around the nucleus is $\hbar$. Applying this rule, the closed loops made by the virtual antifermion and the energy contribute angular momentum $\hbar$. Spin angular momentum is a vector perpendicular to the plane of the antifermion's rotational motion, as shown in Figure 3.08.2. The motion can be on any plane hinged to the fermion's path, right, left, above, or below, and the vector is perpendicular to the plane. Let us observe this vector from the side of the fermion's path. I see only its vertical vector component of the angular momentum. If the loop motion occurs on the planes on both sides of the fermion's path including the direction of observation as shown in Figure 3.08.2 (below), the angular momentum is either straight up or straight down to the observer. Its magnitude is $\pm\hbar$. If the plane of loop motion makes angle ϑ to the horizontal direction of observation as shown in Figure 3.08.2 (above), the angular momentum's vertical component is reduced to $\hbar cos\vartheta$. If $\vartheta = \pi/2$, the component is zero. If I observe the spin only up or down in the vertical direction, I see the average of the angular momentum vector, that is half of $\hbar$, $\hbar/2$, since the average of *cosine* between angle 0 and 90 degrees is 1/2. The vector component is then $\pm\hbar/2$.

This mechanism explains why $\pm\hbar/2$ angular momentum is assigned to the transition edge of the equivalent circuit's waveform, for both a fermion and a boson. The *observed averaged spin* is down if the antifermion's loop of motion is on the observer's side of the fermion path as Figure 3.08.2, and if the loop is on the other side, the spin is up. For spin direction, the right-hand rule applies. Which side and angle ϑ are the antifermion's plane of motion? That depends on the fermion's spin, that is, up or down.

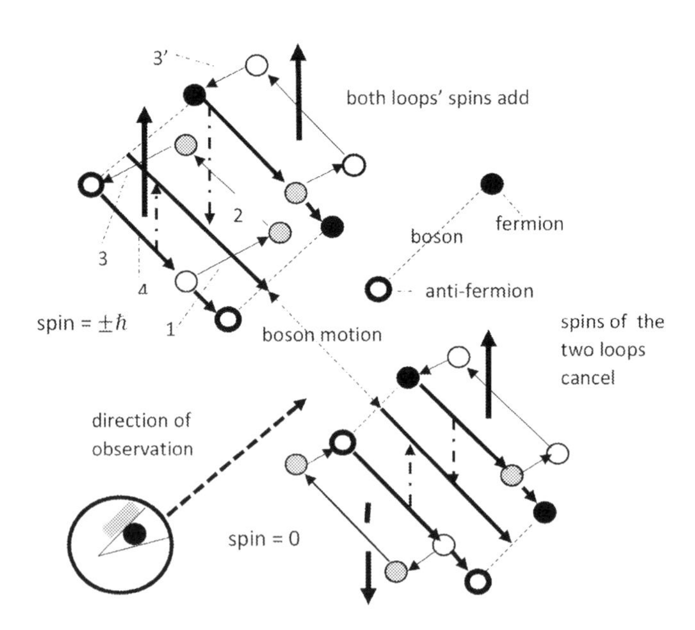

Figure 3.08.4 Component fermion loop structure creating spin of boson

This spin mechanism by the fermion's twisted cycloidal motion explains another feature of the fermion's spin. Why does a single fermion state accommodate two states having opposite spin? A single fermion state actually accommodates two fermions; the one is spin up and the other spin down. This is possible because of the two possible sides of the planes of the associated virtual antifermion's motion to the path of the fermion, one front to the fermion path (spin down) seen by the observer, and the other behind the fermion path (spin up). Let us see the virtual antifermion motion from the direction of fermion motion. If the plane of the motion is always on the left side (observer's side), the spin is down, and if the plane is always on the right side, the spin is up. Thus, a pair of up and down spin states of the fermion emerges. The two spin states fill the single fermion state completely.

A boson consists of a pair of fermion and antifermion. The spin generation mechanism is a combination of the mechanisms of the pair. The fermion-antifermion pair making up the boson is shown by the circles having thick boundary in Figure 3.08.4 (above). There is a pair of two loops, one each for the two edges of the boson's isolated pulse waveform in the equivalent circuit model. Of the two edges of the boson's equivalent circuit waveform, the up-going edge (leading) is associated with the antifermion's loop motion and the down-going edge (trailing) is associated with the fermion's loop motion.

In Figure 3.08.4 (above) the annihilation 3 belongs to the loop that moves the antifermion of the boson, which belongs to the up-going waveform's edge with spin up. The annihilation 3' belongs to the loop that moves the fermion of the boson, that makes the down-going waveform edge with also spin up. The pair of the planes, on which the rotation of the virtual antifermion and the virtual fermion occurs (for the leading and the trailing edges of the boson pulse, respectively) are independently hinged to the line of the boson's direction of motion. The planes can rotate around the direction of the boson motion, as it did in the fermion, and therefore the averaged spin is $\pm\hbar/2$ each.

For boson, at the pair of the up- and the down-going edges of the circuit waveform, spin direction may have four possibilities (up, up), (down, down), (up, down), and (down, up). Spin vectors of Figure 3.08.4 show the *averaged* angular momentum for the two cases, (up, up) (above) and (down, up) (below), along with the plane of rotation of the associated virtual particle that gives the maximum angular momentum. Adding the two contributions, the spin of a boson can be $\pm\hbar$ or 0. Since spin 0 state has two possibilities (up, down) and (down, up), this is a doublet. Thus, spin is the consequence of the *unobservable* associated particle's *twisted cycloidal motion* observed from the side of the particle's path. Therefore spin is universal for all the particles having widely different mass. Since mass of the particle depends only on the particle's state of *translational* motion, independence of spin angular momentum from the mass is an interesting consequence of the rule of angular momentum quantization. Spin is truly a basic character of the elementary particle, much more than the particle's mass. How do you think about this simple mechanism of the mysterious spin? I think this is the model.

3.09 Splitting and Joining Fermion's Path

A fermion's propagation path may split, and the two split paths may join. The simplest case is modeled by the equivalent circuit of Figure 3.09.1, where the path is first split and then the split paths are joined together. This figure shows the upper umbilical cord only. The lower umbilical cord remains unchanged. The probability of existence of the fermion in the pair of paths splits also. By splitting a path into two paths, the probability of finding the fermion is ρ ($0 \le \rho \le 1$) in the main path and $(1-\rho)$ in the side path.

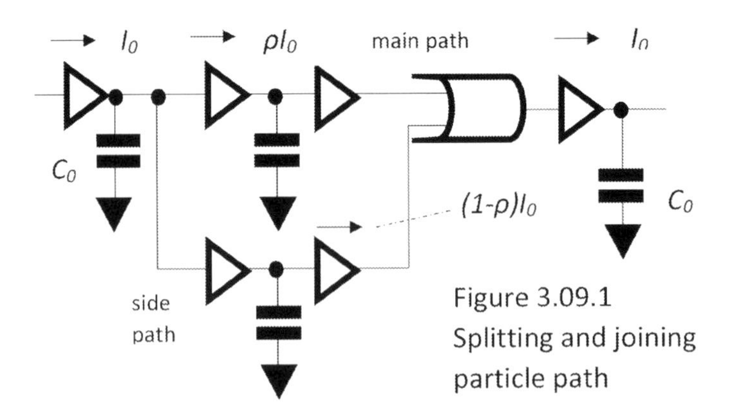

Figure 3.09.1
Splitting and joining particle path

Both of the paths cannot be observed simultaneously; observation of the one path finds or does not find the fermion, and that affects the results of observation of the other path. Observation of a fraction of the fermion never occurs. An elementary fermion represents one bit of information, existence or nonexistence, and if the one bit is split, the split bit cannot specify a fermion definitely anymore. The chance of finding the fermion becomes probabilistic. This is the case in any probabilistic process; in a one-winner lottery, the winner cannot be two partial persons. By splitting and joining, the energy and momentum of the fermion are conserved, and coherence of the pair of fermions' wave functions is maintained. Coherence remains even if the split pair of particle paths are a long distance away from each other. The two paths share a single wave function.

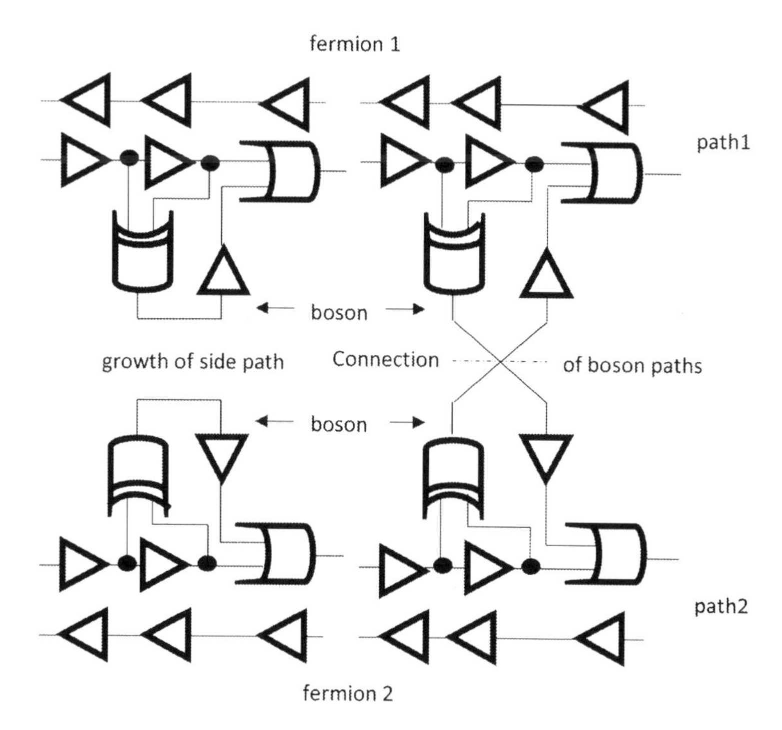

Figure 3.09.2 Boson path formation by the side loop growth

Splitting and joining fermion's path are the mechanism of creating the boson's paths, which transmit the force between the pair of fermions. Figure 3.09.2 shows a pair of fermion paths 1 and 2, one above and the other below the figure. As the paths approach each other closely, the fermions feel the force. Sensing force initiates the boson's activity. The boson's path is then created. Figure 3.09.2 (left) shows how the pair of side loops for the boson path are created. The exclusive OR gate creates a boson (section 3.01), using the delay time of the fermion's path. Since the split boson loop carries some of the fermion's energy away, it is capacitively loaded.

At first each fermion emits a boson and then absorbs it later. This begins to affect the fermions' motion. As the side paths of the pair of fermions come close to each other, the side paths contact, break open, and connect to build a pair of bosons' paths between the pair of fermions, as shown in the right side of Figure 3.09.2. The pair of paths transmit the bosons between the fermions. The paired paths are the bosons' *umbilical cords.* The paths are capacitively loaded, since the bosons carry some energy from one fermion to the other, and vice versa. By the pair of boson paths, the fermions *exchange* bosons. Because of this energy transfer, the interaction takes time. A pair of fermion receptor-launcher emerges after the boson emission and absorption (not shown in Figure 3.09.2).

In Figure 3.09.2 (right), the split path formation and connection have wide spatial flexibility. The boson paths can extend to any direction from the pair of fermion paths, they can bend, stretch, shrink, or entangle to couple the pair of fermions in various ways. This flexibility allows the nuclear force to transmit between any fermion pair and induces all sorts of fermion interaction. Through the pair of paths, the bosons exchange energy between the fermions and keep establishing a new fermion state, such as change of the direction of motion and velocity. The boson's umbilical cord is not static; it moves as the fermion moves and continuously changes shape. This is the basic process of fermion-fermion scattering interaction. It is considered that a real fermion has an umbilical cord and a side tail, and a pair of fermions interact by entangling their side tails, in addition to being connected to their own sources by the umbilical cords.

3.10 How the Character's Charge Transmits Force

Interaction of fermions is exercised by the force working between the character charges of the fermion pair. The source and the recipient of the force are the fermions' substance, their mass or energy, and the go-betweens are the bosons propagating along their umbilical cords connecting the pair of fermions, as discussed in the last section. The pair of fermions that exercise force each other make the bosons by giving away some of their energy. The bosons and their umbilical cords are created by splitting the fermions and their paths, as the means of exchanging their energy. The force depends on the type of fermion and the condition of interaction. The fermion not only carries umbilical cords to its source, but also has a structure on its side, to exercise force with other fermions. They create the invisible network in the three-dimensional space. The quantum vacuum may be considered as being filled with the network of fermions and of connecting bosons. The fermion's *side tail* structure is created by splitting the neutral mother particle when the fermion pair was created.

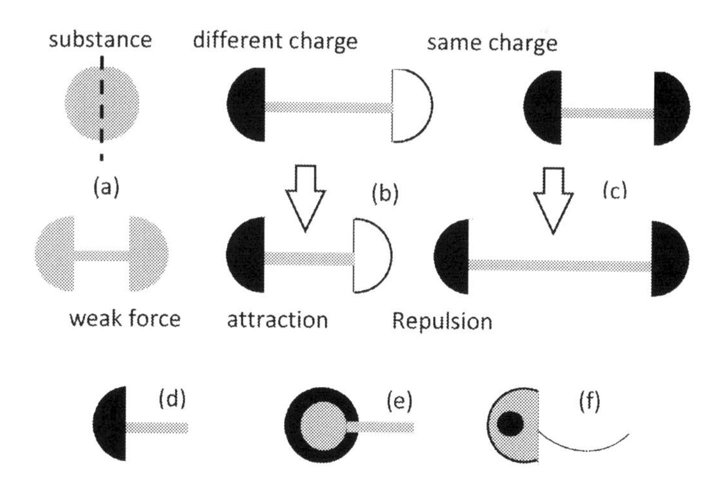

Fig3.10.1 Particle's substance, character's charge, umbilical cord and working force

If the mother particle carries no charge, it may split into the two daughter particles carrying neutral substance, as shown in Figure 3.10.1(a). The force acting between the daughter particles is the weak force of gravity. Why is the force so weak? Gravitation's *character charge* is energy. How do the two bundles of energy exercise attractive force? Material carrying negative energy has never been found, and it is likely that such a material never existed. So the force is always attractive.

The particle-splitting process may go along with sorting the positive and negative charges of the mother particle, as shown in Figure 3.10.1(b). Then the split pair carries the excess energy spent for sorting and separating the charged daughter particles. Therefore the particle pair exercises a stronger attractive force between them than gravitation. The split pieces try to join together to keep the energy low. This is the electrostatic attractive force.

Particle splitting is a sequence of multiple steps. If the mother particle is already charged by previous splitting, the energy becomes lower by further splitting it. Such a daughter particle pair moves away from each other to reduce the energy still further, as shown in Figure 3.10.1(c). The repulsive force emerges from this mechanism. Attractive force is more archaic in nature than repulsive force. This may be the reason why gravitation was the first force that emerged by the first phase transition of the universe as mentioned in the cosmology theory.

Splitting a mother particle is a bidirectional process, of moving positive charge from the daughter particle B to A, and then moving negative charge from the daughter particle A to B. Thus the connection between the split pair of particles is bidirectional, which becomes the boson's umbilical cords that were introduced in the last section. The pair of paths makes up the *side tail* of the fermion to exercise force.

A fermion stretches out its tail to search for the nearby fermion's tail. As such, fermion is a spread-out structure in the space, both to the direction of motion from the source and also to its sides. The tail can freely rotate or has the capability of stretching and shrinking, because its model is the same buffer chain as those of the fermion's umbilical cords. The tail has a dangling end that can connect to the similar end of the tail that approaches from the other fermion. If there are more than two fermions, the connection becomes a complex, entangled net-like structure.

The fermion's tail structure that becomes the umbilical cord of the boson stretches out from the fermion as shown in Figure 3.10.1(d) and then (e). The simplified symbol is shown in Figure 3.10.1(f). The tail of the fermion is a loop, consisting of a pair of go-and-return boson paths at the beginning. As the fermions come close to each other, this dangling tail end is cut open and connects to the tail of any other fermion to create a pair of bosons' path, the boson umbilical cords. In doing so, the bidirectional connections between a pair of fermions is created. The paths move with the fermion's substance from one fermion to the other, thereby creating fermion-fermion interaction. The cord connecting different polarity charges tends to straighten out and shrink as shown in Figure 3.10.2(a). The force exercised by such a cord is attractive. As for the cord connecting the same polarity charge, it straightens out and further stretches, thereby pushing the fermion pair apart, as shown in Figure 3.10.2(b). The force is repulsive.

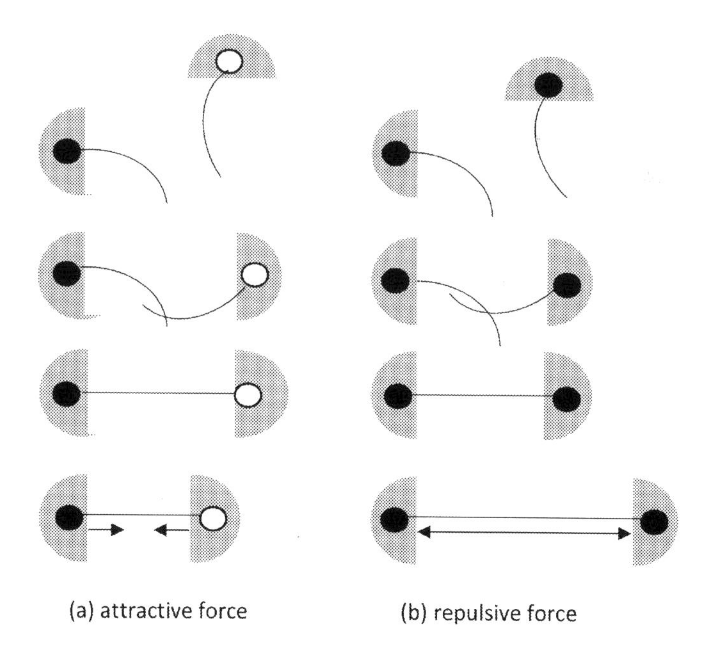

Figure 3.10.2 Entanglement of boson umbilical cord

Cord connection and boson propagation take time, and that is the time of exercising the force and transferring the energy between the pair of fermions. The structure of the tail changes continuously with time, as the pair of fermions move. If the boson umbilical cords become too long, the pair of cords are cut off, and each of them connects to itself to close the loop once again. The original local loop structure reemerges, and that further shrinks to the single propagation path as shown in section 3.09. The pair of fermions are now far apart, and their scattering process is over.

The connection and structure of the boson's umbilical cord shown in Figure 3.10.2 has the classical model of the line of force among the particles. The boson's motion is bidirectional (section 3.09) by the pair of paths, and therefore the interactive force can be expressed as *boson exchange.* Some force of the umbilical cord has limited range (weak nuclear force), and some other forces have infinitely long range (electromagnetic force and gravity). The force is transmitted through the cord at most at the velocity of light, since it is the path of the boson that carries energy, derived from the interacting fermions.

3.11 Entangled Particles

Suppose that a fermion-antifermion pair, such as electron and positron, is created by a common source such as a zero-spin boson, as shown in Figure 3.11.1. They fly away to the right and left, respectively. The pair of fermions are coherent; that is, they share a single wave function. Since the boson from which the pair was created had zero spin, the positron's and electron's spin must cancel, but the spin of each particle is unknown, until the spin of one of the particles is measured. Let the spin of the positron, which moved some distance to the left, be measured.

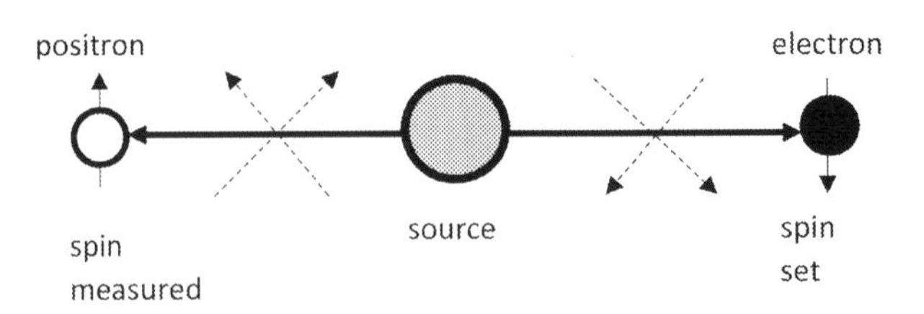

Figure 3.11.1 Quantum entanglement

At the moment of the measurement, the spin of the electron is set such that the sum of the two spins is zero, that is, set opposite to the positron's spin. This effect occurs even if the pair of particles is a long distance apart. This effect, called quantum entanglement, seems to violate the relativistic requirement of locality, that no information carrying signal travels faster than the velocity of light (K. W. Ford, *101 Quantum Questions*, Harvard University Press, 2012). Since probability is not information, this is not surprising to me, but an uneasy feeling remains among physicists, even after the experiment proposed by John Bell has proven that the effect does indeed exist. A lot of books have been written on this subject. Yet this is a direct consequence of the Copenhagen interpretation of the quantum effect, in which I have full confidence. My work on the model of quantum phenomena began to explain this *mysterious* effect by the equivalent circuit model.

Figure 3.11.2 shows the digital equivalent circuit model of this effect. Both particles develop the lower umbilical cord pair between the particles and the common source. The source sends a pair of fermion's (electron) and antifermion's (positron) step functions representing them moving to the right and the left, respectively. The particles move at a velocity less than light velocity, by the capacitor jump mechanism of sections 2.09 and 2.10. At the time of the spin measurement, the charged capacitors are only at the right and the left ends of the respective umbilical cords. The particles there are observable, by measuring their spin.

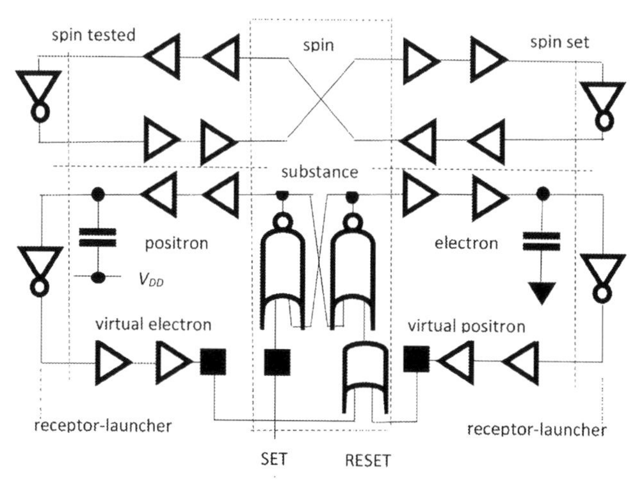

Figure 3.11.2 Equivalent circuit model of quantum entanglement

The spin umbilical cords are the upper pair of the buffer chains. The spin umbilical cords are built along with the substance's cords (section 2.07). The spin up state carries logic HIGH level, and spin down state carries logic LOW level, respectively. The upper left spin cord gives the positron spin, and upper right spin cord gives the electron spin. The paired lower spin cords carry the spin of the virtual antiparticles associated with the positron (left, virtual electron) and with electron (right, virtual positron), respectively. The paired lower virtual particle's spin cord will carry the opposite logic level to the upper spin cord, once its spin levels are set. But the spin levels of both particles immediately after their generation have not been set, and therefore they are unknown.

Let the positron spin be measured at the left end of the spin umbilical cord. The spin measurement detects the positron's substance also. Then a positron receptor-launcher emerges at the left end of the positron's umbilical cord. In Figure 3.11.2, the positron's receptor-launcher is shown simply by an inverter, indicating the receptor's essential logic function. The inverter sends back the positron detection acknowledgment signal via the lower umbilical cord of the positron's associated companion, the virtual electron's, to the particles' common source. The signal resets the latch of the particle pair's source. Then the positron's umbilical cord is set at the logic HIGH level, indicating the empty state of the positron. The positron that traveled the umbilical cord is now sent out from the emerged receptor-launcher to the new state to the left side of the figure (not shown).

The latch of the particle pair's source is reset, and it sends a logic LOW level signal via the substance's umbilical cord of the electron (the upper cord on the right side). Because of the state change, an electron receptor-launcher emerges at the right end of the electron umbilical cord (also shown by an inverter). The electron that reached there is sent out from the electron receptor-launcher to the new state of the right side (not shown). All the processes after positron spin detection take place instantly, since none of the umbilical cord node is capacitor-loaded (section 2.07).

Then what about the state of the spin umbilical cords shown above the substance's umbilical cords? The spin umbilical cords are also a pair of go-and-return buffer chains (section 2.07). Spin umbilical cords are made along with the substance's umbilical cords, and none of its node is capacitor-loaded, since spin does not carry any energy. The spin umbilical cords have crossover at the location of the particle pair's source. From the left side, the spin cord of the virtual electron (the associate of the left-going positron, the lower left spin cord) connects to the right-going electron's spin cord (the upper right spin cord). From the right side, the spin cord of the left-going virtual positron (associate of the right-going electron) connects to the left-going positron's spin cord (the upper left spin cord). This spin cord crossover connection is to assure that the experimental arrangement is left-right symmetrical. This is required, since the particle and antiparticle are symmetrical.

When the particle pair is created, the sum of positron and electron spin is zero, since the mother particle had zero spin. But the spin state of each particle is unknown. The CMOS buffer chain model is able to hold such an indeterminate spin state. In the state, all the PFETs and NFETs of the spin umbilical cord buffers are turned off, and the buffer output nodes are electrically floating. Such nodes' voltages stay somewhere between the logic HIGH and LOW levels. CMOS technology allows such an indefinite state, called the tristate. This is an advantage of using a CMOS equivalent circuit model to simulate quantum effects. During the particle pair motion, the spin umbilical cords extend to the left and the right, while maintaining this indeterminate spin state, tracking the pair of particles' motions.

As the spin of the positron is determined at the left end, its associated companion, the virtual electron's spin, is set opposite to the detected positron spin, to satisfy the spin angular momentum conservation at the left end of the spin umbilical cord. In the lower left spin umbilical cords, the indefinite state changes to a definite state. This definite virtual electron's spin level is transmitted through the virtual electron's spin umbilical cord (the lower spin cord at left) to the source of the particle pair, and crosses over to the right side, to the umbilical cord of the electron's spin cord (the upper spin cord at right). The signal reaches the electron at the right end, and sets the electron spin opposite to the positron spin. There, the electron's companion, the virtual positron's spin, is set opposite to the electron spin, i.e., the same direction as the formerly detected positron spin.

The virtual positron's spin is sent to the particle pair's source, where the virtual positron spin umbilical cord from the right side crosses over to the positron's spin umbilical cord to the left side (the upper spin cord at left), and the spin level reaches the positron's location. This *feedback* mechanism confirms the detected positron spin. This feedback action takes place instantly, since the spin umbilical cord is not capacitor-loaded. Thus, the positron and the electron spins are securely set to opposite directions. The pair of positron and electron carries a definite spin level now, and they become independent from each other. Each of them has its own wave function. The left and right spin receptor-launchers become the sources of the detected positron and electron spins, respectively. The particles departing to the left and the right receptor-launcher have definite spin. Now, the pair of particles are independent, carrying definite spin. After that all the umbilical cords of spin and substance are wiped out.

Since none of the node of the spin umbilical cord is capacitively loaded, the entire process of spin-setting takes place instantly, whatever the distance between the pair of particles may be. This mechanism is symmetrical with respect to positron and electron. If the electron spin is measured, the positron spin is set to the opposite direction instantly. The equivalent circuit model provides a realistic model of nonlocal particle pair spin-setting.

Here I note the important feature of the digital equivalent circuit model once again. The delay time of the buffers that make up the umbilical cords of the substance and the spin does not include relativistic delay time, because the buffer model originated at the phase 2 universe (sections 1.06 and 1.07). Inductance is not included in the circuit model, and therefore the gate operation is nonrelativistic. The probability-setting signal propagates at infinite velocity (section 1.06) through such buffer chains. Quantum entanglement is the definite proof of this crucial nonlocal feature.

The quantum entanglement effect and its model show how similar are the features of the quantum phenomena and of the digital circuit. The digital equivalent circuit does not include inductance, and therefore it works nonrelativistically. Such an electronic equivalent circuit, both linear and digital equivalent circuits, are the proper model of quantum phenomena.

In retrospect, this study was motivated to disprove the popular opinion that human self-consciousness is the quantum effect, and quantum entanglement is the mechanism of ESP, extrasensory perception (*Self-Consciousness, Human Brain as Data Processor*, iUniverse, 2020). The human brain is surely a huge digital circuit that simulates some quantum effects. Several influential authors have claimed that self-consciousness is a quantum effect. I believe that they reached a hasty conclusion by not recognizing that quantum system operation and digital circuit operation have so many similarities. The human brain shows quantum mechanical features, but by itself, it works as the most complex digital circuit ever existed.

3.12 Afterthought

The equivalent circuit model-based study provides a unique viewpoint to basic quantum mechanics. What I discussed in section 1.06 and 1.07 raises a question to the present trend of cosmology, trying to explain the origin of the universe by integration of quantum mechanics and general relativity. I believe that the early quantum universe that could be simulated by triode, resistance, and capacitance was not subjected to a relativistic mode of operation. Relativity is relevant only in the world that includes inductance, and that is the present, phase 3 universe. In the phase 3 universe, the recalibrated parameter values like the buffer drive current and buffer length (section 3.01) assure that no information-carrying signals have velocity greater than light velocity. In the phase 2 universe, where the signal is only probabilistic, light velocity was not the limit of the velocity of a probability-setting signal. I suspect that the theoretical difficulty of the present cosmology originates from reliance on relativity at the beginning of the universe; relativity integrates space and time, and in the early universe, both space and time did not exist in the way assumed by the present, phase 3 universe.

A digital equivalent circuit model of quantum phenomena highlights the role of the capacitor, which carries the particle's energy. I explained this feature in detail in the capacitor jump mechanism (sections 2.09 and 2.10). In the particle motion, *the energy follows where the capacitor goes* (section 2.07). The model distinguishes clearly the role of abstract energy and the realistic energy holder, the capacitor. In classical physics modeling, their close relation appears self-evident and is taken for granted. In my quantum phenomena model, the energy holder capacitor emerges first, and then that is filled by the energy, and various phenomena emerge. If an energy holder does not exist, energy cannot play its role. This is the case even in classical physics. When a nonmagnetic human body is placed in the strong magnetic field of a MRI machine, there is no sense of discomfort to the body. The human body cannot be the holder of magnetic energy.

The quantum vacuum *appears* active. It generates energy δE, during time δt. These quantities satisfy the uncertainty relation $\delta E \delta t = \hbar/2$. The generated energy is eventually taken back, and the energy balance between the quantum vacuum and the physical space is maintained. The apparent energy generated by this mechanism in the physical space is enormous; it is 10^{120} times larger than the energy estimated from astronomical observation (M. Kaku, *Physics of the Impossible*, Doubleday, 2008). This is one of the most remarkable unexplained discrepancies between theory and observation in modern physics. I suppose this much difference may suggest that the quantum vacuum excitation energy does not contribute to astronomical effects at all. What can be the possible reason?

The give-and-take of the quantum vacuum's energy is real, as clearly evidenced, for instance, by generation of an extremely heavy weak boson in the weak nuclear reaction that converts an isolated neutron to a proton. The weak boson's mass is almost 100 times greater than that of the neutron. That much energy to create the mass of a weak boson can be borrowed from the quantum vacuum and then returned. In this case, the boson becomes the energy holder, which is equivalent to the capacitor in the equivalent circuit model.

Yet the quantum vacuum excitation generates energy everywhere, even in the empty space that occupies most of the universe. Then where and what is the holder of the energy? I cannot identify any. In most parts of the universe, quantum vacuum energy has no mechanism to be held and to become effective. Does such energy manifest any physical effect? If most of the quantum vacuum excitation energy is ineffective in creating any astronomical effect, the great disagreement is at least partially resolved. The quantum vacuum energy is like the money in the Federal Reserve bank: until a borrower shows up, the money has no economic effect.

In section 2.06, I quoted that there are authors who believe that there is a *wormhole* of the probability signal transmission in the quantum entanglement effect. I am not convinced of that idea yet, but if that is really the case, there is an interesting possibility; a digital circuit model can be a proper description of the topology of the space created under the state of general relativity. In the very basic level, connection between the spatial points is more important than the actual distance between them in quantum phenomena. This is also the case in digital electronics. Either way, more attention should be directed to find the significance of the digital equivalent circuit model in frontline quantum physics.

In this work, to sum up, relativity is the basic principle of physics of the *present* world. Yet whether it is applicable at the very beginning of the universe, when space and time were quite different, is doubtful. By assuming so, a proper model of quantum mechanics, consisting of resistor and capacitor handling a probability signal, can be built.

CONCLUSION

Quantum physics has been developed from its beginning to the present day by the philosophy that the role of the theory is to identify the quantum system's measurable parameters and to develop the mathematical theory to relate the parameters (P. A. M. Dirac, *The Principles of Quantum Mechanics*, Clarendon Press, 1935). In the twentieth century, the philosophy worked quite successfully. At the beginning of the twenty-first century, however, the theory has gone far ahead, and experimental evidence cannot provide the check. The theory's direct verification is no longer possible by any means available on the earth. Instead of experimental proof, making a convincing model of the theory is one way to get past this difficulty.

Direct experimental checks of the present theoretical frontier such as string theory or loop quantum gravity are, most likely, impossible with the resources available to humankind. The present practice of theorists proceeds by evaluating the mathematical beauty of the theoretical structure, which is, to say the least, not in line with the honorable tradition of natural science since the Renaissance. In this circumstance, the only way to persuade human minds is to create models that directly appeal to human sense. I was motivated by reading popular string theory books, which state that the theory requiring 10-dimensional space-time can never be explained except by going through the theory's mathematical structure. I have seen that there are a few mathematically beautiful theories that have no reality. A proper model resolves this problem. I recognize that the attempt of making models does not entirely belong to physics; it requires involvement of the science of human mind, psychology.

This work is an attempt in this direction, applied only to elementary quantum mechanics. I made several models of the strange quantum phenomena and checked how they appealed to my mind. As such, this work does not belong entirely to the rigorous domain of physics, but neither are my conclusions the stuff of science fiction. My objective is to prompt further research in the basic physics front by using this method and to open a new field of basic research. This is a challenging task. I have as yet no idea how to make a model of any hyperdimensional quantum phenomena.

Before concluding this book, I would like to acknowledge the profound influence of Morris Swadesh on my technical works. My life's works are carried out by a single plan. In the 1960s I was deeply impressed by the glottochronology of Morris Swadesh. His method was able to clarify the relations among the ancestry of native Americans over the entire continent. The basic method was this; 19 percent of basic words used in normal human life, such as *I*, *we*, *bird*, *eye*, and *rain*, about 200 of them, spoken by an isolated tribe changed to different words every thousand years in any part of the world, Europe, Asia and the New World. Swadesh discovered this remarkable feature from what was suggested by radioactive decay of the nucleus, such as uranium changing to lead. The uranium atom becomes the model of a word, the element in linguistics. Methodologically, he *projected* the ideas of linguistics into nuclear physics. By this method, he was able to theorize the New World colonization process into five phases: first, the "lost tribes" now living in the Amazon rain forest; second, Caribbean aboriginals; third, the great Maya tribes arrived; fourth, the great Quechua tribes of South America arrived; and last, the Inuits in the arctic moved in. How much we learned about the New World antiquity by this method can never be overstated. In reality, I learned how to create a new science for the first time in my life.

Then a question came to me: if my lifelong interest in the electronic circuit theory's models were *projected* into other scientific branches, what would come out? Following this plan, I projected electronic circuit theory, first to thermodynamics (*Theory of CMOS Digital Circuits and Circuit Failures*, Princeton University Press, 1992), second to electrodynamics (*High-Speed Digital Circuits*, Addison-Wesley, 1996), third to Newtonian mechanics (*The

Dynamics of Digital Excitation, Kluwer Academic, 1997), fourth to psychology (*Neuron Circuits, Electronic Circuits and Self-Consciousness*, Vantage press, 2009; *Self-Consciousness, the Hidden Internal State of Digital Circuits*, iUniverse, 2013; *Self-Consciousness, Human Brain as Data Processor*, iUniverse, 2020), and fifth to quantum mechanics (*Equivalent Circuit Model of Quantum Mechanics*, iUniverse), that is, this book. I am presently considering my next project, which will be to project the circuit theory to chemistry. I am still struggling along the direction set by the great linguist Morris Swadesh's inspiration.

Printed in the United States
by Baker & Taylor Publisher Services